NEW SCIENCE LIBRARY

presents traditional topics from a modern perspective, particularly those associated with the hard sciences—physics, biology, and medicine—and those of the human sciences—psychology, sociology, and philosophy.

The aim of this series is the enrichment of both the scientific and spiritual view of the world through their mutual dialogue and exchange.

New Science Library is an imprint of Shambhala Publications.

General editor/Ken Wilber

Perceiving Ordinary Magic

Science and Intuitive Wisdom

Jeremy W. Hayward

NEW SCIENCE LIBRARY
Shambhala
Boulder & London
1984

NEW SCIENCE LIBRARY
An imprint of Shambhala Publications, Inc.
Boulder, CO 80306-0271

Distributed in the United States by Random House
and in Canada by Random House of Canada Ltd.
Distributed in the United Kingdom by Routledge & Kegan Paul Ltd,
London and Henley-on-Thames.
Printed in the United States of America

Library of Congress Cataloging in Publication Data
Hayward, Jeremy.
   Perceiving ordinary magic.
   (New science library)
   1. Perception.   2. Neurophysiology.   3. Meditation.
I. Title.   II. Series.
BF311.H412 1984   153.7   84-5481
ISBN 0-87773-297-3 (pbk.)
ISBN 0-394-72704-5 (Random House: pbk.)

Figure 2 reproduced from *Space-Perception and the Philosophy of Science*, by Patrick A.
Heelan, © 1983, published by the University of California Press.
Figures 9 and 10 reproduced from *States of Mind*, by Jonathan Miller, © 1983, published by
Pantheon Books.
Figure 11 reproduced from *Maps of the Mind*, © 1981 Mitchell Beazley Publishers Limited.
Text © Charles Hampden-Turner 1981. Illustrations ©Mitchell Beazley Publishers
Limited 1981.

Design/Eje Wray
Cover illustration/Jurgen Peters
Typesetting/The Type Galley/Boulder CO/Linotron Palatino
Printing/Fairfield Graphics/Fairfield PA
Cover printing/Strine Printing/York PA

Dedicated

To my mother and father,
in deep appreciation

To my teacher, Chögyam Trungpa, Rinpoche,
in unlimited devotion and gratitude.

Contents

Preface

The Shambhala teachings of the Path of the Warrior, which form the philosophical and practical basis of this book, are a nonreligious presentation for modern times of the possibility of awakening to complete perception of this world, free from aggression, and to the gentle and fearless action which is part of this perception. The possibility of such awakening is based, as in the Buddhist tradition, on the practice of mindfulness and awareness meditation. The Shambhala and Buddhist teachings point to an utterly fresh and delightful perception of this ordinary world, which is available to everyone when we see beyond our presuppositions and the partial and narrow views to which we habitually and often unknowingly cling. Meditation practice provides the means to uncover our preconceptions to the deepest and most intimate level.

The history of science is the history of continual revelation of partial truths which have often been clung to as the whole or the absolute truth. Nowadays these partial truths are, indeed, often themselves created by science. Nevertheless, the role of science seems to lie far more in its disclosure of partiality than in any steady progress toward some supposed "truth." If we take the insights of science personally, it, too, can throw light on the beliefs that constrain and entrap us. Partial truths have narrowed and limited our perception of and participation in the ordinary world of nature in all its vastness and ineffability. When we begin to free ourselves from these limitations, by seeing them clearly as such, then we can discover a fresh perception of this world purified of preconception and narrow belief. We may, then, perceive ordinary magic—a world which is vast and unconditioned, yet full of power, value, and natural order.

If a broad view is to be found which can accommodate the insights of both science and the practice of meditation, then it must begin with a thorough reexamination of the most fundamental presuppositions on which science is founded, an endeavor which Alfred North Whitehead carried out in the 1930s with very great success. Whitehead's process philosophy remained a quiet undercurrent in philosophical circles for several decades, but now as positivist and behaviorist types of philosophy fade away in usefulness, process philosophy stands out as one of the few trends of thought in the West that can bring back some sanity and heart. Process philosophy grounds science not on high abstraction, but on the most immediate moment-by-moment experience of human perception.

This book, then, is about perception. I would like to invite the reader to take a journey on which we might, together, discover some of our own preconceptions and deeply hidden assumptions about our world. Thus we might occasionally glimpse, beyond them, immeasurable and inconceivable nature, which is at the same time not separate from the ordinary world we live in. Scientific insight into our deep-rooted presuppositions can be of help to us in our spiritual journey through this world, and at the same time, the practice of meditation can be grounded in a thorough scientific understanding of the human body and mind.

The book is intended for the general reader: little knowledge of physics or biology is assumed, and no mathematics is used. I have tried to keep the language as simple as possible, not out of depreciation of the apparent complexity and subtlety of the issues, but because what I wish to convey is essentially simple. It may also be of interest, and perhaps even helpful, to those who, having an appreciation for the great heritage of science and at the same time an interest in the practice of meditation, wonder how these two could possibly be referring to the same world. The suggestion that I make in this book is that, yes, they are dealing with precisely the same world, this one and only lovable world.

My heartfelt thanks go to all the friends who read and criticized various revisions of the manuscript, and who have over the years contributed many of the ideas in this book; to the faculty of the Cognitive Science Program at Naropa Institute in the summers of 1979 through 1981; to the faculty of the Buddhist Studies Department of Naropa Institute; and to the many Buddhist friends around the world. I would very especially like to thank Dr. Reginald Ray and Dr. Francisco Varela for their extensive and invaluable criticism, and most of all my wife, Karen Hayward, whose editorial work and detailed comment at each stage of the manuscript contributed essentially to its clarity and accessibility. Mr. Samuel Bercholz, friend and publisher, first suggested that I write this book, and without his unfailing encouragement and gentle guidance throughout the process, it would remain even now just a suggestion. Finally, I would like to thank my teacher in the Buddhist and Shambhala traditions, Vajracarya, the Venerable Chögyam Trungpa, Rinpoche. Words are certainly inadequate to express the gratitude I feel toward him. Whatever small glimpse of understanding may shine through these pages is due to his guidance and teaching, and to his patience and humor.

In addition, I would like to acknowledge with appreciation the hospitality of the staff of Rocky Mountain Dharma Center during a retreat at which a major part of the writing was done; also the OSCO company and Flatirons Surveying for use of their word processors. And, of course, my great appreciation and thanks go to Rene Evenson and Michelle deRaismes King for their untiring and cheerful work in typing. Beyond this, I would like to thank Mrs. King, my administrative assistant, for her years of kindness and unwavering support.

Perceiving Ordinary Magic

Part One
Ground: Unconditioned Nature

1 Unconditioned Goodness

It is a late September evening. Shadows have become long across the brilliant brown and yellow meadow, still radiating the warmth of an autumn day. Wind whines through a crack in the wall of the cabin. On the other side of the gentle valley pine trees bend, and a strong wind roars in the distance. A dark gray cloud moves swiftly to cover the evening sun. A chipmunk pauses a moment by a rock, then scurries into the bushes. Big, warm raindrops fall. Many insects are still busy on a circle of charred rocks. The air has a full and peaceful quality, yet there is an ominous, threatening edge to it.

The World Turned Upside Down

As we turn to the evening news, or read the newspaper and the penetrating analyses of our world today, we begin to feel anxiety and despair, anger and frustration. We are warned of possible nuclear catastrophe; of imminent ecological collapse; of the pending death of our species. Acid rain pollutes the air; thousands of people a day are slaughtered in dozens of wars around the world; heart disease and cancer, stress and mental illness are increasing; our best students are losing heart in their studies and committing suicide or dropping out of school. These warnings are urgent. It is often difficult to acknowledge the enormity of the danger confronting the world today.

Then there is another side to our present situation: a sense of respect for individual human life; increased communication so that we begin to feel part of a worldwide family; appreciation of the tremendous variety and joy of human cultures throughout the world; the fantastic possibilities offered by computer technology; the possibility of actually overcoming starvation throughout the world; the emergence from superstition of large numbers of people—not just the elite—through general education.

It seems as though humanity altogether could be just about to step into heroic freshness and a genuinely joyful existence on this planet. Yet we feel almost overwhelmed by fear, threat, anxiety, and despair.

There is a widespread feeling that we are on a razor's edge, and that we are presented with a wide variety of both sensible and fanciful suggestions of how to cross this razor's edge to achieve a more harmonious society. Generally speaking, these suggestions all tend to take for granted the biological and historical causes and conditions which brought about this present society. Rather than deeply examine this conditioning, they accept it and try to find solutions to its inevitable consequences by clinging to one particular side—the righteousness of the past, the optimism of the future, or the burden of the present.

The Moral Majority in the United States is the most obvious and extreme example of the attempt to hold onto or return to values and methods that have proved successful up to a point in the past, although it is not clear that the methods they advocate have ever really provided an adequate basis for social relationships. Nevertheless, their wish for clear, incontrovertible guidelines to good behavior, and for everyone to follow these guidelines, has its roots in the Puritan sense of duty and perhaps even the Crusader's ethic. But there are more subtle ways of trying to bring about stability by holding onto the way we were: the structures of political systems, school curricula, business methods, economic systems, all necessarily have an inherently conservative tendency, providing those stable forms by which a society recognizes itself from generation to generation. Because these forms are familiar, because generations have learnt to live within them, they become references to hold onto when the ground of our civilization begins to shift. The Paedeia Proposal,[1] a much-heralded report of a study by high-level educational theorists and practitioners, proposes that the solution to the problems in our schools is to persuade all children, no matter what their talent or interests, through the same curriculum based essentially on absorbing the great ideas of our past. The policies of many governments in the world today, that of Reagan in the United States and Thatcher in Great Britain, for example, are based on a call to reaffirm the old values: our family, our church, our nation, and our God.

The contrary tendency to conservatism is futurism. Some writers speak of a growing number of people, an underground network or "conspiracy," who are bringing about a "new age." These people, in every realm of human activity—medicine, physics, neurobiology, therapy, economics, and so on—are making discoveries that, it is enthusiastically believed, will bring about total revolution in the way we live.

Futurist Alvin Toffler writes, "A powerful tide is surging across much of the world today creating a new, often bizarre environment in which to work and play. Today's changes are not independent of each other. Nor are they random. They are in fact parts of a much larger phenomenon: the death of industrialism and the rise of a new civilization." Toffler speaks of three waves of civilization in human history: The first wave, lasting thousands of years, was agricultural. The second wave, lasting just three hundred years, was mechanical-industrial. We stand now at the transition between the second and third waves. Toffler refers to the bewilderment, anxiety, and paralysis widespread among us as a state of future shock brought about by the rapidity with which this transition is happening. Toffler, however, moves quickly away from contemplating this state of shock of transition to promote hope for the future. "Many of the same conditions that produce today's greatest perils also open fascinating new potentials." Thus, if we can quickly and smoothly bring about this new age (based nevertheless on "many of the same conditions" of the past age), we will redeem humanity.[2]

John Naisbitt, in *Megatrends: Ten New Directions Transforming Our Lives*, highlights some of these new potentials: transitions from centralization to decentralization, from industrial society to information society, from hierarchies to networking, from institutional help to self-help, from forced technology to high technology combined with a high level of human touch, from representative democracy to participatory democracy.[3] Marilyn Ferguson brings it all together in the phrase "Aquarian Conspiracy"—the revolution of humanity against the oppression of mechanism, reductionism, industrialization, and massive bureaucracy.[4] There is almost some implication that this revolution is happening inevitably, with little effort on the part of most of us.

In the middle, between conservatism and futurism, is the pragmatism, rationality, and common sense of liberal humanism, which does not expect too much and acknowledges pluralism and the diversity of human nature. Through open discussion of all points of view we can at least learn to tolerate each other. And perhaps, if we learn to act and talk at the same time, we will be able slowly to eradicate the barriers that separate us. This is the approach of trying to live in the present and accept it as it appears without too much dwelling on the glories and traditions of the past, and without hoping too much for rapid progress, least of all fundamental change. It is a mature shouldering of the sometimes-solemn burden of being human.

So we have the three alternatives of returning to past beliefs and institutions, expectation of revolutionary change within our lifetime, or a rational and limited acceptance of the present. Each of these attitudes has its own positive view of humanity and a genuine wish to bring about something better. Yet the struggle between these viewpoints has been going on for generations. In fact, one finds essentially the same struggle over and over again throughout history and across culture.

Each viewpoint is based on its own version of good and bad. Each sees certain forces for good in the world and in individuals, and certain forces for bad. Each feels that there is a struggle between these forces and that we are somehow buffeted between them and must make a choice, presumably for the good. So the struggle continues between good and evil, right and wrong, upstairs and downstairs, past and future. So long as we cling to these partial views, based on particular conditions of culture, history, and nature, the struggle will not end. Partial views are inherently irresolute, insecure, and exclusive. They have to be maintained and defended and therefore lead inexorably to aggression and struggle on both the personal and the social level. It is the main theme of this book that there can be no fundamental change through any of these ways.

Living Teachings

In order to bring about such change we must go to a deeper, a more fundamental level of understanding. This understanding is always available as a quantum leap out of these reference points to an entirely different view. This is the realization that the inherent nature of human mind and heart, of human existence altogether, is unconditioned and therefore fundamentally, basically good. Beyond all philosophies of good and evil, of righteousness and sin, of morality and immorality—beyond all theories of human development and human behavior, all potential theories of conservatism, futurism, and liberalism—the human being has a basic unconditioned nature. This realization is always fresh because it is based on immediate experience rather than on stale philosophizing, argumentation, and logic, or on wishful thinking.

This simple discovery is primordial; it is independent of the history of race and of culture. It is the essence of the teachings of many people, religious or secular, who have set humanity on a fresh course: Socrates, Jesus, Confucius, Mencius, Lao-tzu, Buddha, George Fox who was the founder of Quakerism, and many others. In this century we might mention Ramana Maharshi, an Indian sage who died in 1948; G. I. Gurdjieff, a Russian who taught in Russia, France, and America in the first half of the century; Krishnamurti, uncompromising and world-renowned proclaimer of the unconditioned nature of man and of the urgency of overcoming narrowness and fixed beliefs of every kind; and Chögyam Trungpa, scholar and meditation master trained in the Vajrayana Buddhism of Tibet, who is founder and teacher of the Shambhala Path of Warriorship.[5] The unconditioned goodness of human nature has been proclaimed, exemplified, and taught by genuine leaders of people over and over again. Very often such teachings arise at times just like the present, when the forces of conservatism and change have ceased to work together and have become sharply opposed. When Socrates lived in Athens, the sense of security and grandeur was being questioned by wandering teachers such as Socrates himself. Jesus lived in a time of subjugation and turmoil amongst the Jewish community. Confucius was born into a period of chaos and warring forces in China. Buddha lived at a time when the indigenous cultural and religious forces of the people of Maghada (a province of the Indian subcontinent, now in Nepal) were being challenged by the conquering culture of the believers in Brahma and the caste system.

Because it speaks of, or points to, *unconditioned* nature, the teaching, although simple, is extremely profound. It is like a time bomb which, once it has entered an individual or a society, is inevitably bound to bring about fundamental change. The possibility of fundamental change today comes not from trying to recreate the past, or to project the future (which is then in any case only a recreation of the past), but to recognize, now, the transition we are in. At this very time there may be an opening in our minds and in our cultures for genuine freshness to come in. This energizing and powerful possibility comes from the fact that we have thoroughly seen through old belief systems and forms of conditioning, not that we are already beginning to find new beliefs to adopt and fight for. If we can recognize and not avoid the anxiety and fear that arise as our belief systems fail, and if we can stay with this fear and not try to jump too soon onto a bandwagon of new beliefs, then there is a possibility of opening into a vast view.

Wherever this teaching appears, it appears in a form exactly appropriate for the time, place, environment, and people. It is spoken in the language of the time, be it Greek, Aramaic, Chinese, Maghadan, or English, and it uses the cultural norms and basic cosmological beliefs of the people as its form. In other words, it is presented as a direct, living experience. As times and circumstances change, the *form* in which such teachings are presented becomes fixed, itself the force of conservatism and rigidity. Meanwhile the living teaching has to take on new forms. The form of the teaching is preserved as a philosophical or religious system which is by then divorced from immediate experience. The people become confused, following a form which, while supposedly arising from direct experience, in fact leads them away from it. Therefore, in order to understand the teaching of basic, unconditioned nature and its discovery and expression in the world, we must relate it to our own experience. It will not be of very much benefit merely to take it as yet another philosophy.

Unconditioned Goodness

Now let us examine further the meaning of unconditioned goodness. I have said that it is the fundamental nature of our existence altogether. It is the very ground which is human existence. We might have various theories that this fundamental basic ground of existence is "matter" or "mind" and so on, but when we go further back in our experience than all these theories, we find that this ground is something else much more immediate. At this level, goodness has nothing at all to do with morality, with "good" as opposed to "bad," or with being "for" or "against," or with any discrimination of this kind. It is more at the feeling level that we begin to understand why we use the term "goodness." We could perhaps say beneficence or wholesomeness. All of these terms have a similar sense but carry with them philosophical connotations: beneficence is mostly used nowadays in a religious context, and wholesomeness is related to "holistic," which has a rather specific and narrow meaning. So we have the term goodness, which is to say that, at the most fundamental level, everything works, everything fits and functions together. We have eyes that see, and our eyes see red as red, yellow as yellow. We have ears that hear, noses that smell, tongues that taste, skin that touches. There is food that nourishes, water that cleanses, fire that warms, air to breathe, impressions to know. The body transforms all of these, functioning with tremendous accuracy, even in sickness. There is heartbeat, pulse, and nervous energy. The human being is capable of making love, of touching and stroking another tenderly, of caring for his or her young, of becoming angry and overcoming aggression. The human being has a fundamental appreciation for his existence in the universe. Beyond this, the universe itself is workable: nuclei and electrons form atoms, which form molecules, which form DNA, organs, organisms, and so on. Mountains, rivers, valleys, animal species, plants all function together. The earth itself adapts to changes as if it were a living organism. There is the sun which warms, the atmosphere which encloses and brings circulation, rain which moistens, the seasons which bring freshness and rebirth. There are the stars which may well be teeming with more forms of life and intelligence.

The human being has the capability to see and feel this goodness, not taking it for granted, on a very simple level. Our appreciation that we can see red as red and tend our young is appreciation of the workability, the goodness, of all basic existence.

The universe is not a static, monolithic mechanism. It is constantly, unpredictably changing, constantly evolving and self-creating. Because there is change, there is life and creativity. The heart of the universe, and therefore of our own nature, is change, which is life. Because our fundamental nature is unconditioned, we always have the potential to overcome partiality and fixed beliefs, which are the source of fundamental aggression. From this point of view the unconditioned in our experience IS goodness, wholesomeness, healthiness. The unconditioned, because it is free of all limitation and boundary, free from all particular condition, is without any exclusivity, without any sense of separation of one thing from another. Unconditioned nature is not a particular thing, a specific separate object of sense perception or thought. It is not a God, it is not Oneness, nor is it Nothing. It is also not solemn. Because it is unconditioned, it cannot be limited by any finite verbal description. Therefore we cannot answer a question of the form "What *is* this unconditioned?" It is without specific name or form. Even the term "unconditioned" does not adequately describe it. Descriptions of what is, of conditioned reality, simple or grandiose, holistic or reductionistic, may turn us in the direction in which we might recognize our unconditioned nature. But that recognition is up to each one of us individually and momentarily. Nevertheless, such recognition brings about a transformed and utterly fresh perception of this very world in which we live. It is the possibility of uncovering this new perception which this book proposes.

Transforming Perception

We will be discussing a transformed perception of this ordinary world, a perception which is very personal, real, and heartfelt. Although we will be reviewing apparently abstract ideas of space, time, matter, and life, I would ask the reader always to take these things personally. "Space" refers to the actual room in which you now sit, and to the space between the atoms of your own body and brain. "Time" refers to this very moment as you read, "matter" to the solid world around you and in you, and "life" to that sense of your own beating heart and flickering thoughts that is so personal to you. It is this world about which we will be speaking and this world in which we may discover our unconditioned nature of goodness. Because it is an inherent aspect of our nature—of our mind, heart, and body—this discovery is an unfolding process, rather than something foreign put into us. Such unfolding, like growing up, takes discipline and a training process. How, then, are we to discover this unconditioned dimension to our experience?

Perhaps you are sitting at the kitchen table. You have finished breakfast, washed the dishes, and are ready to go out for an appointment. But you are not in a hurry, and you sit there looking out of the window at the street. It is a warm day with high clouds. People are walking down the sidewalk in bright summer clothes—blue suits, a red and yellow dress, a white shirt. Some are moving briskly, almost urgently, others are strolling. A horn toots, a vendor cries his wares, a child laughs. Suddenly your eye catches a flag flapping in the breeze on the roof across the street. For a moment your mind rests with the flag. A sudden flash of freshness, almost of joy, enters your state of being. Your mind seems suddenly fresh, and open, your body relaxed and warm. There is a quality of lightness, humor, and almost timelessness. It is very simple. Yet for the remainder of the day you feel utterly refreshed for no good reason. This is a flavor of basic goodness.

Unconditioned goodness is said to be the most immediate, intimate, and all-pervading aspect of our everyday experience. It is always there; it does not come and go. Our recognition of it may be fleeting, but it does not depend on any particular circumstances. We may be in a very mundane situation, such as I have just described, or in a more extreme situation. A sudden glimpse of unconditioned goodness is always possible. Traditionally such realizations are said to be most likely to occur spontaneously at a moment of sudden surprise, or they occur as a result of the practice of meditation. Still, the glimpse of openness can come along at any moment. Such moments are a first step. If they are acknowledged and allowed to enter our state of being, then gradually our life begins to open out; we begin to cheer up and feel a natural wholesomeness and dignity. If we do not acknowledge these simple experiences, then they are forgotten and our life continues along its particular and familiar grooves. Discovering and acknowledging such moments constitute a journey of opening that needs some discipline and some training.

As Chögyam Trungpa says: "Discovering real goodness comes from appreciating very simple experiences. We are not talking about how good it feels to make a million dollars or finally graduate from college or buy a new house, but we are speaking here of the basic goodness of being alive—which does not depend on our accomplishments or fulfilling our desires. We experience glimpses of goodness all the time, but we often fail to acknowledge them. When we see a bright color, we are witnessing our inherent goodness. When we hear a beautiful sound, we are hearing our own basic goodness. When we step out of the shower, we feel fresh and clean, and when we walk out of a stuffy room, we appreciate the sudden whiff of fresh air. These events may take a fraction of a second, but they are real experiences of goodness. They happen to us all the time, but usually we ignore them as mundane or purely coincidental."[6]

The simplicity and naturalness of the discovery of unconditioned goodness need to be emphasized again and again because, although we may never have studied philosophy or been particularly religious, we nevertheless are very much influenced by the philosophical and religious attitudes of our culture. Scholars, theologians, materialists, and occultists have made a very big deal out of "reality." They have tried to be convinced, and to convince us, that any sense of connection with something beyond our own individuality, any connection with the profound basis of our existence, is either impossible, or only for the chosen, or meaningless.

Unconditioned Nature Dawns in the West

Questions about unconditioned reality are certainly not unknown in the world of Western scholarship and in fact go back to the very beginnings of Western speculations—before Socrates. But the idea of the unconditioned as void, nothing, *nihil* has always won the day in the cultural mainstream. This notion became an implicit part of Western culture until this century. The effect of this attitude toward the unconditioned quality of experience comes out very clearly in the history of the encounter of the West with Buddhist doctrine.[7] Such encounters were sporadic until the eighteenth century and the establishment of the British East India Company. When in 1784 William Jones began to translate Mahayana Buddhist texts and send them back to England, Buddhism was met with a mixture of horror and revulsion and dismissed either as merely a movement of sociological reform, or as demon worship. Most fiercely attacked and misunderstood was the notion of the unconditioned realm, nirvana. Since this was taken to mean *nothing* and was often translated as "void," Buddhism was taken to be some kind of pessimistic, world-denying religion of self-destruction.

These kinds of misinterpretations continued until the beginning of this century when, for the first time, living representatives of the Buddhist meditative tradition began to visit the West. The first such visit was by Soyen Roshi to the World Parliament of Religion in Chicago in 1893. His simple, practical statement impressed people, especially by its contrast with their preconception that Buddhism was world-denying. Soyen Roshi left behind his young translator, D. T. Suzuki, whose vitality, kindness, and gentle scholarliness did much to begin to reverse Western misunderstanding of Buddhism.

In the 1920s and 1930s, other Zen Buddhist teachers quietly came to the United States and began to meet in small groups with the intelligentsia and artists. These teachers provided living examples of the possibility of awakening in this very world and gave instruction in the paradoxical Koan practice in which a student ponders an apparently absurd statement until his conceptual mind realizes its own limitations. These ideas began to filter into the culture, but were nevertheless taken very intellectually.

When we look back at this period we can see that something profound was beginning to break through. The discovery of relativity and of quantum theory had begun to put into question the belief that science was telling the ultimate truth about the world or indeed that such truth could be told at all; the logical positivists of the Vienna Circle were questioning the meaning of truth altogether; mathematicians were stunned by Gödel's demonstration that a complete and consistent system of logic is impossible; musicians, painters, and dancers were beginning to experiment with entirely new ways of listening, seeing, and moving. There was an attempt to break through the heavy sense of rationality, of righteousness and certainty, received from the nineteenth century, a turning toward this ordinary world for inspiration and for the discovery of abstract principles of harmony and disharmony, form and chaos. There was a sense of newness and daring. But at the same time something was lost. When the shackles of tradition were thrown over, the profundity and wisdom of that tradition were also often thrown out.

In the forties and fifties, avant-garde artists—in particular, John Cage—and Beat poets Kerouac, Ginsberg, Whalen, and others were profoundly influenced by the Buddhist sense of the unconditioned "nothingness," as it was most frequently translated in those days. As Calvin Tomkins reports, "Cage believes that the world is changing rapidly and more drastically than most people realize. Cage insists that the true function of art in our time is to open up the minds and hearts of contemporary men and women to the immensity of these changes, in order that they may be able 'to wake up to the very life' they are living in the modern world."[8]

There was, however, an overemphasis on the emptiness aspect of Buddhism, to the neglect of the sense of natural harmony of, for example, Japanese Zen art. This is understandable, since the object was to show the extent to which music, painting, poetry, and so on had been conceptualized and to break through to nonconceptualized, spontaneous art. However, "nonconceptual" was often interpreted as merely formless or even antiform. Most unfortunate of all, the "emptiness of mind" of Zen was often regarded as meaning that we should try to destroy our intellect; when drugs came along in the sixties, some young people tragically succeeded in doing just this. But this has nothing to do with "emptiness of mind," or discovery of the unconditioned. In fact, sharpness of discriminating intellect is an essential instrument of such discovery. These artists were able to bring their audience to an intellectual realization of the extent of its conceptualization and expectation about music, poetry, and so on, but a further step was needed: a discipline by which the unconditioned dimension could be directly discovered, nourished, and brought out in one's daily life.

In the late sixties, the time bomb which had been lit in the early part of the century finally exploded. No longer just the fringe intelligentsia or artists, but a significantly large percentage of the population of the United States, began to believe that there was something profoundly lacking in their lives. Daniel Yankelovich, a Harvard-trained philosopher and pollster, puts the figure at close to 17 percent.[9] In addition, according to Yankelovitch, another 63 percent, although not strongly affected by these changes, were nevertheless affected by them in some form. Thus the spiritual supermarket and the search for self-fulfillment of the seventies began. This was a period of self-indulgence and also of openness. Many people rejected their families and jobs, causing suffering to themselves and others in a search for their "self." Thus the seventies were rightfully referred to as the "Me-First Decade." Yet there was also a genuine search for a more profound appreciation of life. Yankelovich has shown that many of the same people who in the seventies were desperately searching for "self-fulfillment" have, in the eighties, begun to look beyond this for lasting commitments and opportunities to genuinely give to others. Yankelovich has dubbed the eighties the "Decade of Commitment." He proposes that the decades of the seventies and eighties are analogous to an adolescent who must first break with his parents and their values in order later on to rediscover those same values and be able to practice them with sincerity rather than empty imitation.

During the seventies there was a genuine aspiration to extend the boundaries of life beyond the narrow limits set by materialism and behaviorism, and beyond the disease orientation of therapists. There was tremendous striving to bring about a better world. Psychedelic drug experiences indicated that things may not be as we ordinarily perceive them to be, and this was a profound shock to many unsuspecting people. In a search for a larger context for understanding and appreciating the world, many people turned to Eastern traditions, particularly Hinduism and Buddhism. This period was like a fantastic feast: numerous "new religions" and self-proclaimed gurus popped up and disappeared again. People dressed in yellow robes, red robes, and black robes. There was chanting, dancing, meditating, praying, vegetarianism, fasting, yoga, sexual freedom, and communal living. It was a wonderful experiment; some people became attached to the foreign forms as a new crutch, something new to believe in; some tried a few experiments and gave up; and some found a genuine path.

The Discipline of Meditation

One of the most profound and lasting consequences of this period was the introduction by living masters of various traditions of the practice of sitting meditation. Theravadan teachers such as Munindra and Mahasi Sayadaw; Mahayana Zen Buddhists Suzuki Roshi, Maezumi Roshi, and many others; Vajrayana Buddhist Chögyam Trungpa all taught essentially the same discipline: sit mindfully, and put this mindfulness into action in daily life. It is so simple, yet it is the most powerful discipline for turning us directly toward our basic, unconditioned nature.

The power of the discipline does not depend on its association with any particular tradition, religious or otherwise. The practice of sitting, or mindfulness-awareness meditation, is a simple human statement of recognition of who we are. It is uniquely human and something that many people come to when they have exhausted the possibility of trying to solve their problems by chasing thoughts or trying to fulfill every desire. The practice of meditation is a way of resting the mind so that it may rediscover its origin and its harmonious integration with body and world. It is a powerful tool for seeing clearly the nature of mind and perception and for training our perception, and is as scientific as the use of a microscope to examine tissue. We shall see in later chapters that there is a scientific basis for the process of training our minds. It is a training process which is natural and necessary if we are to know fully our human possibilities.

That great American psychologist at the turn of the century, William James, said, "The faculty of voluntarily bringing back a wandering attention, over and over again, is the very root of judgement, character and will. No one is *compos sui* if he have it not. An education which should improve this faculty would be the education *par excellence*. But it is easier to define this ideal than to give practical direction for bringing it about."[10] Unlike in William James's day, practical direction for bringing this about is nowadays publicly available in the Western world. The simplest form of such practical direction is bare attention or sitting meditation.

Bare attention or sitting meditation is a primary practice of many great contemplative traditions, both religious and secular, particularly Buddhism, neo-Confucianism, and Taoism and some contemplative orders of Christianity. It is also the primary practice of the teaching of Shambhala, a nonsectarian, nonreligious synthesis of Buddhism and neo-Confucianism for modern times. The Shambhala teachings provide a path for the awakening of our genuine nature in the ordinary world. It has been presented by Chögyam Trungpa for the past ten years and forms the practical and philosophical basis of this book.

It is important to distinguish and clarify what is meant by meditation here. We can broadly divide practices that go under the general title "meditation" into two types which are rather distinct in their background assumptions, in their techniques, and in their effect.[11] Because these two types have not been clearly distinguished, there has been a great deal of confusion on the subject of "meditation" in the West. These two types could be called trance type and openness or access type. The basic presupposition of the trance type is that there exists another realm of experience which is more significant in a variety of ways than the ordinary realm we exist in—"inward" and also "higher." Therefore the technique of the trance type of meditation is to try to close off one's awareness of the senses and even of thoughts, to turn inward.

The presuppositions of access-type meditation are that it does not matter whether or not other realms exist; that perception of the ordinary world is clouded due to unexamined preconception and biological predispositions; that, by training of the attention to be more directly connected with the senses and thought process and more responsive to sensory and conceptual details, it is possible to clarify the cloudiness. The technique of access-type meditation is training of attention and identification with thought and perception in sitting practice and in everyday-life activities.

I should add that there are times when concentration-type practices can be helpful in overcoming particular conceptual obstacles, and therefore use of these practices within the overall context of development of openness is not ruled out. Nevertheless I want to point out the difference between these two types of practice and to clarify that in this book, when I refer to meditation, bare attention, or mindfulness awareness, I mean the openness, access type of practice associated, as Pelletier says, with the "doctrine of being open to and appreciative of every aspect of the phenomenal world."[12]

The simplest way to discover our unconditioned nature for ourselves is, then, the discipline of sitting meditation, or mindfulness-awareness meditation, which has nothing to do with trying to attain "higher" states, "alternate" states, or union with something greater. Meditation, here, also does not mean meditating *on* or thinking about a particular subject. It simply is a way to become attentive to our thoughts, emotions, perceptions, bodily sensations, and environment, so that who and what we are begin to be clearly and precisely seen. We sit in a relaxed and upright posture, with straight spine, open chest, hands resting naturally on the thighs. To take such a posture already expresses the genuine dignity of being human. To remain in that posture during the ups and downs of our thought and emotional processes expresses the fundamental confidence of trusting in unconditional goodness. The eyes are open with soft gaze, slightly down, and we take the same attitude to the other senses—open but not fixed or harshly striving to experience something. As we sit there, we allow our minds to identify with the outgoing breath, to go out with it, and then to return to be attentive to the posture as the breath comes in. As thoughts, feelings, and physical sensations begin to pop up, we note them and let them be as they are, not trying to push them away, or holding onto them and indulging them. We begin to become mindful of the precise details of our thought and perceptual processes and also aware of the relationship between them. A thought or feeling arises, and then it goes away. Where it arises from and whence it goes, who can say? But occasionally we might catch a glimpse of non-thought, of open mind. A glimpse can be tremendously refreshing. It is such a relief to realize that we can afford to let go of our conceptualizing process altogether. Such a glimpse of our basic nature of unconditioned goodness brings with it a sense of gentleness and tenderness toward ourselves.

Such experiences are probably quite familiar to many of us, so familiar in fact that we usually tend to dismiss them as being, although nice, of no particular significance. Yet such simple moments *are* of significance. They are the footprints of and the pointers to our unconditioned nature. They are the gaps in our tightly organized and controlled lives through which a glimpse of brilliance and clarity is seen.

To acknowledge our unconditioned nature and the possibility of uncovering it through a discipline of meditation is to form the basis of a society in which human relationship and political structures are based on genuineness, on knowing who we are, on the natural empathy of tenderness and gentleness. The primary value of such a meditative discipline is, ultimately, to bring about such a society in order to help others. Nevertheless, in order to help others it is first necessary to at least begin to see our own individual conditioning, our self-imposed and inherited limitations and prejudices. The uncovering of our conditioning will be the main topic of this book, especially our conditioning as it is revealed, reestablished, and confirmed by the scientific tradition. We will try, however, not to lose sight of our purpose in doing this, which is to discover the gaps in our conditioning through which the unconditioned might shine through and to train our insight so that we might directly apprehend this unconditioned nature. In the next chapter, we will look more deeply at what is meant by unconditioned nature, and in Chapter 16, we will discuss the details of mindfulness-awareness meditation.

2 Three Natures and Insight

This morning a rainbow appeared in the apparently clear blue sky. It lingered for fifteen minutes or so and then vanished. Some time later a large flock of small birds flew over in a V-formation. Several of the birds kept sweeping out of the formation, played together for a while or settled on a tree, and then returned to the flock. The V-formation remained unbroken. Toward the late afternoon a beam of sunshine made a warm column of air in the chill evening. A swarm of gnats darted back and forth in the last few moments of sunshine.

Our bodies are like the swarm of gnats, or that V-formation, perhaps even the rainbow. They are made up of conglomerations of atoms, every one of which will be replaced as the years go by. It has been roughly calculated that every breath of air we breathe contains a few atoms that have been breathed by every person in the history of the planet, from Socrates to Genghis Khan to Einstein to Hitler, as well as all the billions of unknowns. Thus the form of our body may seem more substantial than the stuff it is made up of. And the apparent continuity of this form may lead us to believe that there is a self or "I" that is "more real" than the body. On the other hand, the atoms appear to be hard little things. But then in turn these real, hard little things are simply the form of the way electrons and nuclei move around each other . . . formations of formations of formations. Considerations like this have led many philosophers to conclude or perhaps hope that this world is altogether an illusion and that there is something *more* real above, beyond, underlying, or in it. Others conclude or perhaps fear that there is nothing real at all.

On the other hand, the pragmatist will say, "Listen, my friend, just wait till you hit a patch of ice at sixty mph and skid off the road into a tree. *Then* we will see what is real and what is not." This is a very good point. We usually, and quite sensibly, take our own body as the commonsense reference point for reality. If you see a chair and wonder whether you are hallucinating it, try sitting on it. Beyond this, the earth itself becomes our reference point for the reality of our body: if we wonder whether we are dreaming, we try kicking a rock. If we lose this reference point of the reality of body, the result can be psychosis. If we deny the reality of other people's bodies the result can be terrible neglect of hunger, poverty, and disease. Yet all these very real things are formations of formations of formations . . .

Eternalism and Nihilism: Extreme Views of Reality

The tendency to hold onto an extreme and therefore partial view about reality can be divided into two general styles known as eternalism and nihilism. Eternalism is the view that there *is* a substantial reality, that there exists some permanent unchanging substance or thing which is absolutely real and from which everything else arises. This might be termed God or Brahma, spirit or self, or matter or energy or field. We might run the danger of thinking that unconditioned goodness also is such a permanent, existing *thing*.

This is the commonsense, naive view that, of course, the world is real and exists. But since most things in everyday life are not permanent, there must be something permanent underlying them which explains them and exists eternally. This view also comes in more subtle forms: the tendency to see all religions as speaking of "the same thing," the One; the tendency to find an underlying substantial reality in the "fields" of the new physics; the tendency to think of laws of economics or political structures or human behavior as having some fixed, objective, immutable character, which explains these things and provides us with moral laws.

Nihilism takes the opposite view: that there is fundamentally nothing, that the final, absolute truth is that all is chance and beyond that there is nothing. Apparent existence, and the laws of physics and biology and so on, are purely accidental. Therefore, how we behave, how we live our lives, make no difference whatsoever. Each moment of our lives we are faced with a choice which seems very important to us, yet, in fact, is utterly meaningless. If you have the kindness to care that someone else is in pain, that is your business and it is nice to help him. In the short run such behavior might bring some positive reaction. So you see cause and effect and realize that you can live a more comfortable life by behaving decently toward others. But at the absolute level it means nothing whether you help them or kick them in the face.

Each individual is isolated in a bleak, empty, meaningless universe, exists for a short while, and at the moment of death is totally extinguished. We might decide to try to seek constant pleasure and entertainment, or not to bother. We might just decide to end it now. Or we might feel a certain heroism and decide that, purely on the basis of objective logic, we will be kind to others. We might even turn it into a kind of religion: the fact that we cannot know anything is itself the absolute truth. But at bottom, life is a meaningless and stupid accident.

In personal experience these two fundamental attitudes manifest themselves as different ways of trying to avoid looking at one's basic nature, different ways of managing to feel secure and definite about one's life. The eternalist tendency is to believe in one's own eternal existence. This is reinforced with habits and habitual patterns of thinking and behavior: enjoying certain particular foods and not wanting to venture into a new restaurant; always going to the same spot for one's vacation; listening to certain types of music and despising other types; having definite opinions about one's friends and acquaintances, liking some and taking offense at the habits of others; getting up at a certain time, eating at a certain time, feeling particular moods at particular times, and being very disturbed if things do not occur on time. Of course our habitual patterns are by no means all narrow ones. One may be quite a famous painter and with a definite and well-known style; someone else may find that courting physical fear in hang-gliding or mountaineering is a way he can keep confirming himself, and so on. All of these little ways of structuring one's life give one a feeling of definite existence.

However, sometimes holding onto habitual belief in our existence becomes too much. This is the situation with the nihilist, who may feel hopeless and lost, without friends and constantly in danger of losing everything, even his sanity. Yet the nihilist clings even to his nihilism and loneliness, perhaps feeling that he alone is willing to face the harsh reality. Or perhaps he simply dwells in doubt and despair.

So in all these little ways we spin a web, a cocoon, around ourselves. The cocoon becomes nice and snug and comfortable because it is very familiar. We know every little corner of our life; we can even write poetry about it. We may also have ideas about the "great mystery" which religions speak of, which gives our cocoon an especial sense of security: we can worship the great mystery outside of it and feel good about that. The cocoon is safe, bounded, claustrophobic, and a little stale. We settle into it and live our lives.

The Conditioned World and Our Beliefs About It

In the previous chapter I pointed out the possibility that there is another dimension to our experience, the unconditioned dimension, and that the rediscovery of this dimension can enrich our lives. I said that the unconditioned refers to experience, not to some philosophical or abstract ideal, and that the practice of sitting meditation points us toward this experience and provides us with a path by which this rediscovery may gradually unfold. In this way there gradually evolves a new perception of the world as already good and pure in itself. In this new perception we realize that we ourselves are not separate from the nature and power of the phenomenal world. In this chapter we will try to discover what is meant by the unconditioned in relation to the ordinary reality of our physical, biological, and conceptual world.

In order to begin to discover our unconditioned nature we must look at our conditioning, and at the way we take our habitual patterns for granted as "the truth" and let our life be dictated by them. Our conditioning itself is not a problem. As we will see, our conditioning provides the boundaries, the constraints, within which we live and function in the world. In fact, so long as we have a body these constraints are very specific and very deep—far deeper than many of us would like to believe. No, it is not our conditioning but our *beliefs* that are the problem. That is to say that the problem is the beliefs that are laid on top of and interwoven with the conditioning as interpretations of it. These are not examined, but taken as "the truth," meaning "the final truth" or "the absolute truth." Beliefs provide a sense of certainty about *something*, but that sense then becomes an obstacle to further opening and inquisitiveness. To begin with a very simple example: your friend John becomes angry, shouts at you, and strikes you. Although the physical pain is negligible, you in turn become angry at John and hit him back. Why? Is it because you assume that inside that body that hit you there is a "John" and that this "John" willfully hit you? It was the willful anger of this "John" that caused your anger, not the physical pain of being hit. But is there in fact a "John"? And was not his anger a product of something else as much as was your angry response to him? We take a great deal for granted in our ordinary experience of life. And this "taking for granted" constitutes our beliefs, which in turn become habitual patterns. As Alfred North Whitehead wrote in *Science and the Modern World*: "In every age the common interpretation of the world of things is controlled by some scheme of unchallenged and unsuspected presupposition: and the mind of any individual, however little he may think of himself to be in sympathy with his contemporaries, is not an insulated compartment, but more like one continuous medium—the circumambient atmosphere of his place and time."[1]

Our conditioning itself can function at many levels, from the deepest genetic predispositions of our organism to our inherited and learned ideological or cultural norms. As examples of genetic predispositions, we might take perception of the world as a container with objects in it; perception of certain physical situations such as height as threatening; or fear of loud, unexpected, and unrecognizable noises. Examples of cultural constraints might be the division of labor between men and women in some cultures that results in the woman's spending more time in the house tending children and keeping house; cultural activities and institutions that attempt to recognize the fundamental equality of each person, and so on. Cultural constraints include all of the "background" behavior by which a culture interprets itself and humankind. Such background behavior may not be known, and to try to know it is to change it already. Later on, we will be carefully examining these conditionings and also the way they interact with each other. For the time being, the important point is that we should not think that our conditioning itself is a problem. For example, we are not saying that it is either wrong or right to be afraid of heights, or to jump at loud noises, simply that it is a natural disposition of our organism to do such a thing. Nor are we saying that it is wrong or right that, under certain conditions, women have been more helpful to themselves and their group when they have adopted a home-tending role; nor that it is wrong or right that cultural institutions recognize the common humanity of everyone. However such cultural norms often come to be believed as "the truth" about "the world." They may be accompanied by slogans such as "Woman's place is in the home" or "All men are equal." Nor are we saying that pointing to the uncon-ditioned dimension of experience means that we can be entirely free of such conditioning. This would be to suggest that we can begin to fly or to start seeing red as blue, which is not the case. Nevertheless, the depth of our conditioning together with the extent to which we take it for granted and *believe* in it as the final truth *is* a problem if we wish to live genuinely and fully.

The Three Natures

Buddhist practice and thought have provided further insight into the relation between (a) the unconditioned, (b) the conditioned, and (c) beliefs. In the Buddhist tradition these three are known as the *three natures* and are variously translated from Sanskrit.[2] Some sample translations would be (a) ideally absolute, or absolutely accomplished, *nature;* (b) relative, or other-dependent, *nature;* and (c) notional-conceptual, or imaginary, *nature*. Thus the first *nature* has the quality of being complete (ideal, accomplished) and without relative reference (absolute). The second *nature* has the quality of being relative to or dependent on something other than itself (other-dependent), the product of cause and effect. The third *nature* is imaginary or conceptual in the sense that our concepts about the world are taken to be real and to stand for real things. These three *natures* correspond to what we have called unconditioned, conditioned, and belief.

Understanding the interrelationship between these three is to understand that to state the reality of the unconditioned is not by any means to deny the relative reality of the conditioned, the interdependent world of cause-and-effect relationships. It is, however, to deny the reality of the imagined or conceptual *nature*. To put it very simply, if we remove from our perception of this ordinary world the imagined or conceptual *nature* (c), then the relative *nature* (b) stands clearly before us and is seen to be not separate from the unconditioned *nature* (a). Or, to say exactly the same thing another way, if we see our beliefs and concepts *as* beliefs and concepts, as partial truths, then the conditioned as such is seen to be not other than the unconditioned. The unconditioned is the inherent purity and immaculateness of the world; the conditioned is its inherent structure. What we have called our cocoon consists of this entire web of the imagined or conceptual *nature*, our beliefs about the world. Furthermore, it is said that we live almost entirely within this imagined realm.

It should be clear that the understanding of the three natures implies that the Buddhist insight avoids the extremes of eternalism and nihilism.[3] In particular, Buddhism does not take the nihilistic view that "this world is an illusion." Unfortunately, such misinterpretation has been widespread in Western writings, both scholarly and popular, since the first introduction of Buddhism to the West. It was in response to these two extremes that Asanga, realizing that the problem was one of *our knowledge* of reality rather than one of reality as such, proposed the three natures. As Janice Willis says, "For him [Asanga], voidness [unconditioned] is grasped 'rightly' by that one who sees with the Middle Path mode of viewing, i.e. who neither exaggerates nor minimizes reality as it really is. He neither denies nor affirms *in toto*. Rather, he sees what he sees as 'just that' and he knows that it is possible for a thing to exist in such a way that it is neither totally existent nor totally non-existent. He views all things as being neither of these two extremes. Consequently, he comes to judge the essential nature of all things as being, in fact, inexpressible."[4]

Let us look at these three *natures* in relation to a particular situation. Two human organisms sit in a field. In the distance is a tree. One says to the other "I see a tree." The situation itself is a product of cause and effect: the two humans arrived in the field as a result of the intercourse of their parents and their whole life history up to that moment. Likewise the tree is there as a result of an entire web of prior and immediate causes. The ability of the organism to perceive and to use language is also a product of cause and effect. All of this is relative, dependent *nature*. But when one person says to the other, *tree*, he is using a symbol which applies not just to that particular tree in front of him but to all trees, and he is using this symbol to convey meaning to someone else. Then the word *tree* begins to be taken as the real thing. We lose touch with the connections of the actual tree to its environ-ment: its roots go deep into the earth and draw sustenance from it; a branch is half dead and decaying where it touches the earth, blending imperceptibly with the earth; the leaves exchange gases with the atmosphere and receive sunlight from it; a seed which a moment ago was part of the tree is now lying on the ground and already initiating the processes by which it will become a seedling. We begin to regard the tree as separate from everything, just as the word *tree* is separate and distinct from all other words. We begin to associate with the tree all kinds of conscious or unconscious qualities coming not from our perception of the actual tree but from our web of connotations of the word *tree*. Finally, the word *I* begins to take on a real existence even though it was used only momentarily in the field to refer to the specific fact that its organism was perceiving a tree.

Likewise, if we say "I am very angry," the idea "angry" begins to become more real than the energy of anger itself, and we may continue to respond to our idea of being angry long after the bare energy has changed into something else. We live a great deal of our life in this conceptual realm. It is the conceptualizing, discriminating, imagining function which separates isolated things and then conceives of their real existence or imagines their connections. When we see through this conceptualizing process, when we see it as just that—a concep-tualizing process—and begin to look directly and closely at it, then we see the interconnectedness of the entire phenomenal world. Everything is seen to have no ultimately independent existence. This is true even of the apparent distinctions between "I" and that which "I" perceives. It is not that we do not see a tree, or experience a certain quality of energy (angry); it is that this tree and this anger are realized as dependent on everything else including the seer and the experiencer. This then is the relative or dependent nature.

As another illustration of the three natures: Alfred Korzybski is famous for having said, "The map is not the territory."[5] This is often taken to mean that the word *tree* is not the tree itself; that is, the conceptual nature is not the relative nature. In other words, the circles and squares on a map are not *themselves* towns and cities; they are just circles and squares. The contour lines are not mountains; they are contour lines. However, we can go further than this. When, after bending over a map trying to locate the direction of some mountain, say Mount Kanchenjunga, you look up in the right direction and there it is, you say, "Ah, there it is." But you do not see only Mount Kanchenjunga. You see also its relation to other mountains, the space around them, and the space between yourself and the mountain. We begin to realize that the mountain itself is not separate from the space around it and its relationship to other mountains, to ourselves, and to the earth and sky. Thus the mountain is nonexistent as a *separate* entity. When we take away the conceptual, imaginary *nature* by not looking at the map but looking at the real world, then we see not just the separate things in the world but their total interconnectedness with each other and with space. To see the mountain as it is, having no separate existence, to see the interconnectedness of mountain, sky, earth, and observer, is to see the conditioned relative nature. Realizing this, one may discover unconditioned nature, which is inexpressible.

Any aspect of relative nature, which is some aspect of this ordinary world free of its conceptualizations, is real always in reference to the other aspects of the world from which it is separated; that is why it is called relative. Unconditioned nature does not take its reality by reference to anything else at all, not even the concepts of existence or nonexistence, cause and effect, realism and idealism, transcendence and immanence, and so on. All of these are part of the conceptual or imaginary *nature* (even though philosophers have often taken them rather seriously). Unconditional nature is sometimes known as "That" or "What is." It is just that, what is, before thought discriminates form. "That" cannot be defined positively, since any description is necessarily partial and relative. "That" shines through all of our activities and perceptions. It is the brilliance and purity and spaciousness of colors, sounds, smells, and so on, as they are before we have a concept of them. The term "unconditioned" is preferable to "absolute" since the latter term has so many theistic connotations—the "One," the "underlying reality," that which is *more* real or the "only reality." These connotations are not intended here. The unconditioned

here is not different from the phenomenal world and is not *more* real
than this world, as it is in itself beyond imagination and concept. It is
only more real than our imagination about this world. By this I do not
mean only the obvious fantasies and daydreams that we all have, or
the subtle misinterpretations we continually make—thinking we are
sick when we are not, and so on. As I have already pointed out, we do
not even see something so simple as a tree just as it is. Our perception
of the tree is narrowed and dulled by our web of associations and
preconceptions connected with the idea of "tree." In a later chapter, as
we discuss the relation between language and reality, the depth of this
conceptual-imaginary world will become more clear.

Intellect and Intuition

I said that descriptions of unconditioned nature cannot fully
correspond to it. Such descriptions become part of the conceptual
nature and thus can lead us further and further into precisely that web
which we seek to see beyond. This does not mean, however, that we
cannot, in immediate experience, directly come across unconditioned
nature.

The present moment, as it is most immediately experienced, is
undivided, unconditioned, and unlimited. Everything is possible;
there is tremendous scope available. Before we perceive or think
anything, there is this felt or intuited background of unbounded
possibilities. Against that background we make distinctions. We
distinguish hot and cold; loud and soft; red, white, green; like and
dislike. And from there we expand immensely. We could say that the
possibility of making distinctions, of differentiating this from that, is
also limitless. George Spencer-Brown in *The Laws of Form* examines this
process of distinction; there is a first impulse to discriminate, to which
the next distinction is connected, and the next, until we arrive at the
whole of science, engineering, psychotherapy, the arts, and so on. "A
universe comes into being when a space is severed or taken apart. The
skin of a living organism cuts off an outside from an inside. So does the
circumference of a circle in a plane. By tracing the way we represent
such a severance, we can begin to reconstruct, with an accuracy and
coverage that appear almost uncanny, the basic forms underlying
linguistic, mathematical, physical and biological science, and can
begin to see how the familiar laws of our own experience follow
inexorably from the original act of severance."[6]

Thus there are two basic aspects to our experience: we have the
bare, unconditioned, and unfragmented moment, and within that is

differentiation, fragmentation. As a first step we could speak of these two aspects as corresponding to intuition, or intuitive insight, and intellect. Intuition or intuitive insight is what is immediately felt; it is an esthetic sensing of the whole moment. Intellect conceptualizes, and thinks about this and that. Intuitive insight and intellect are always co-present; there can't be any experience in which there is purely one or the other. But there can be exaggeration or imbalance.

If we exaggerate intellect, we emphasize quantity, separation, analysis, and abstraction. If we exaggerate intuition, we deny differences, and we experience a lack of clarity and precision—a lack of reality testing. There could be imbalance either way. Although there may have been societies in which there was an imbalance in favor of intuition, and a denial of intellect, the imbalance we experience in contemporary society, and which seems to be the most common in recent Western history, is the overemphasis of intellect. Actually we find that it is a rather gross form of intellect, which stops short of its potential to be sharp, extremely precise, and to make every possible distinction. In Buddhist traditions that potential is described as a double-edged sword that is constantly cutting, constantly dividing that which can be divided, constantly clarifying. It is double-edged because it clarifies even itself. But in contemporary society the process doesn't go far enough, in a sense; we tend to make certain distinctions and then hold onto them as an end, as "the truth," as "reality." That is, rather than fully know conditioned nature in all its details, which is intellect's true function, it stops short at partial views, creating beliefs to which it clings as "truth." As physicist David Bohm says, "Since our thought is pervaded with differences and distinctions, it follows that such a habit leads us to look on these as real divisions, so that the world is then seen and experienced as actually broken up into fragments."[7] Bohm is here describing the imaginary nature.

When the discriminating, analyzing function of intellect is overemphasized, we do not experience the fullness and wholeness of our lives. And it is in this fullness that the good of unconditioned goodness is discovered. But there is a possibility to restore the balance, a possibility to rediscover the nonseparateness of intuition and intellect, of nonthought and thought.

Each moment of one's experience has, then, two interconnected aspects: the nonconceptual interconnectedness of everything, directly and intuitively felt (the unconditioned nature); and the forms arising in awareness which are discriminated within the absolute (the relative or conditioned nature). Now we see that if we take these forms as real, independent of their basis in absolute interrelatedness, then they give rise to the conceptual or imaginary nature. The unconditioned and relative are equally real. The imaginary is imaginary. We might add here that from this point of view the so-called higher realms beyond this ordinary world, such as the "subtle" and "causal" realms of certain mystical traditions, would also have a conceptual or imaginary nature. Thus it is entirely unnecessary to go through such realms which because of their seductive nature are possibly dangerous sidetracks. It is possible to discover unconditioned nature right here in this very world.

While sitting practice points us straight on the path toward discovery of unconditioned goodness, it is clear from the previous discussion that there is a complementary aspect to this path which is to examine and appreciate directly the nature of our conditioning and the conceptual beliefs that we project onto that conditioning. Thus we may gradually unravel our web of beliefs and open the way to direct perception. Fundamentally, the path of training and discovery is extremely simple, direct, and uncomplicated. But that is not to say that it is effortless. The reason that the journey requires effort is that our imaginary-conceptual beliefs form layer after layer, like onionskins. As we peel off layer after layer, we come closer and closer to our most cherished beliefs, beliefs that we may not even realize we have and act by. Even if we do realize we have them, we may take them for granted as absolute truth, instead of realizing their relative nature. For example, consider your own views concerning space, time, perception, mind, and individuality or personhood. All of these are partial views and therefore, as we have seen, cannot be absolute. Yet, whatever our unexamined views on these things are, we usually take them for granted and depend on them as our ultimate reference points for reality and for our responses to reality: our actions, thoughts, moods, and so on. When we see that particular partial and relative views have been taken to be "ultimate truth," then we become free to continue our journey.

Rigpa: Intuitive Insight

The two aspects of our journey, that of direct perception of the unconditioned and that of perceiving and appreciating as such our conceptualizations about reality, are what I have called intuitive insights and intellect. We should be quite clear that intuitive insight is a faculty in man which is capable of direct perception of the unconditioned. It is not quite correct to say "perception" here, since in the normal sense, this would imply a sense of oneself, or a self-consciousness separate from what is being perceived. In this case we are talking about unconditioned perception in which there is no separate sense of self. "There is in man a faculty which is capable of discovering the ultimate [unconditioned] and in order to discern the ultimate man must in some way partake of it. The faculty capable of discerning it is rigpa [intuitive insight]."[8] Although *rigpa* is itself without reflection or theme—that is to say, it is before thought and nonconceptual—it is that which bestows value or meaning on our perception. This property is what is being referred to in Chögyam Trungpa's phrase, "First thought, best thought," which seems to allude to that naked glimpse, no matter how brief, which occurs in our perception before thought about it arises. Perhaps we could say that when one "rigpas" the unconditioned one *is* unconditioned, and therefore there is no self-consciousness at all. For this reason we are not so interested in the development of "self-consciousness" which is fashionable these days. Rather, *rigpa* is spoken of as nonreferential awareness, that is, awareness which does not refer back to self or to anything else, awareness which has no reference point at all: no "watcher," no "witness," no "editor," no checking back of any kind.

Rigpa and intellect are complementary and, at the accomplished level of training, not separate. They interchange back and forth; intellect sees the imaginary nature of our conceptualized beliefs and therefore appreciates the relative nature of our conditioning, thus opening the way for a glimpse of direct perception, *rigpa*. Then, at a further level of conditioning, the whole process is repeated again. Thus the journey continues until all conceptualized belief is seen through and the unconditioned nature is fully experienced.

As we open each moment to intuitive insight, there is freshness and delight, a fundamental cheering up. There is a sense of being relieved of a tremendous burden: our burden of narrow belief. As Guenther says, "It [*rigpa*]is a cognition whose mood is one of unchanging bliss because the fragmentation into a past I may mourn because it is gone, into a future I may dread, and into a present in which I am involved instead of making a decisive choice has been transcended."[9] Although such glimpses may be rare and brief, we realize that *rigpa* is always there (unchanging) like the sun, because it appears when, momentarily, the clouds of depression caused by self-imposed limitation and narrow beliefs are parted. And since, as Guenther says, our being partakes in some way of the nature of the unconditioned, it is inevitable that we will discover it. It should be clear that this journey is not an ego-centered one. In fact, the discovery of our unconditioned nature is the discovery of egolessness. It is the discovery that the idea of a separate self is also part of the imaginary nature.

As Trungpa Rinpoche has said of the discovery of unconditioned goodness on the path of warriorship of the teaching of Shambhala: "By relating with the ordinary conditions of your life, you might make a shocking discovery. While drinking your cup of tea, you might discover that you are drinking tea in a vacuum. In fact, *you* are not even drinking the tea. The hollowness of space is drinking tea. So while doing any little ordinary thing, that reference point might bring an experience of non-reference point. When you put on your pants or your skirt, you might find that you are dressing up space. When you put on your make-up, you might discover that you are putting cosmetics on space. You are beautifying space, pure nothingness.

"In the ordinary sense, we think of space as something vacant or dead. But in this case, space is a vast world that has capabilities of absorbing, acknowledging and accommodating. You can put cosmetics on it, drink tea with it, eat cookies with it, polish your shoes in it. Something is there. But ironically, if you look into it, you can't find anything. If you try to put your finger on it, you find that you don't even have a finger to put! That is the primordial nature of basic goodness, and it is that nature which allows a human being to become a warrior, to become the warrior of all warriors.

"The warrior, fundamentally, is someone who is not afraid of space. The coward lives in constant terror of space. When the coward is alone in the forest and doesn't hear a sound, he thinks there is a ghost lurking somewhere. In the silence he begins to bring up all kinds of monsters and demons in his mind. The coward is afraid of darkness because he can't see anything. He is afraid of silence because he can't hear anything. Cowardice is turning the unconditional into a situation of fear by inventing reference points, or conditions, of all kinds. But for the warrior, unconditionality does not have to be conditioned or limited. It does not have to be qualified as either positive or negative, but it can just be neutral—as it is."[10]

In the next chapter we will look at some of the major ideas concerning the nature of language and its relation to reality that have developed in the past half century. As we discussed in this chapter, misunderstanding the role of language is a fundamental obstacle to clear perception. Language is so intimate. It is difficult to separate ourselves from it, to hear words as simply sounds and to realize that the meaning lies not in the sounds, but in that to which, beyond themselves, these sounds are pointing.

3 Language as Metaphor

The logician had left his family at the end of August to spend the autumn at a retreat place. At the end of September the aspen leaves had turned to brilliant gold, the days were warm, and the evenings were chill enough for him to light the wood stove. He invited his wife and four-year-old daughter to drive up and spend a Saturday with him. Perhaps because of the intensity of the retreat, the beauty of the surroundings, the spontaneity of the visit, or the fact that they had not seen each other for a while, as well as their affection, it was a delightful, joy-filled day.

As they drove away down the dirt road, at the end of the day, he waved until he could see the car no more. Tears came to his eyes and the feeling in his heart can barely be expressed.

> The crickets cry
> With the quelling cold of night
> Autumn hastens on
> And gradually they seem to falter
> The voices traveling away.—*Saigyo*

> Loneliness
> The essential colour of a beauty
> Not to be defined
> Over the dark evergreens, the dusk
> That gathers on far autumn hills.—*Jakuren*[1]

Small incidents such as this one, which each of us experiences many times in our life, raise a question: What is the relation between what we experience and the language we use to describe that experience?

The Correspondence Theory of Truth

In this chapter we will investigate the relation between language and what is, beyond language. We will examine whether there could be a direct correspondence between language and what is. In particular we will question the rather absurd notion that what cannot be described cannot be known.

At school a child begins to learn about facts, which are what actually happens, and opinions, which are just things people believe. He learns that if you say something that corresponds to a fact, you are telling the truth; if not, it is false. He learns that the world is full of things and facts are about things. He learns what smaller parts things are made of, how one thing relates to other things and what things are like. For example, he might learn about the parts of a car, how these parts are put together to make a car, and what different kinds of cars there are. Or he might learn about flowers: their different colors and shapes, their different parts, pistils and stamens, pollen, and so on, and how bees come along and, in collecting honey from the flower, rub pollen from another flower onto it. Or he might learn about atoms.

As he gets older he finds a girlfriend and she says to him, "I love you." This arouses in him a peculiar mixture of excitement and doubt which he cannot put a name to and he says, "Is that true? Do you really *mean* that?" She replies, "Yes." At that point he decides to call his feeling "love" and says, "I love you, too." When he gets older still he begins to search for "The Truth."

All of this involves the belief that the words we utter are in direct one-to-one correspondence with what is "out there," that is, outside of our mind. Every sentence we utter is then believed to stand for what is out there, that is, the things and the relations between them. So every such sentence is either true or false, depending on whether it does or does not correspond to what is in fact "out there." For example, "That tree is green" (said while the speaker points to a tree) is true if that particular tree is, in fact, green. But we can go even further than this and say that the sentence "Trees are green" is true if, in fact, trees *are* green. This view of truth as the correspondence between our thoughts and what actually is the case goes back at least as far as Aristotle, but it is also what most people believe.

These seemingly rather simplistic thoughts have been dignified with the name "the correspondence theory of truth." It is, more or less, what we take for granted about the relation between language and reality. It seems fairly obviously correct when we are talking about whether or not trees are green. But over the generations we have also taken other things for granted. For example, we have taken for granted that "I love you" or "I hate you" also refers to a "fact" in the real world: a real relationship denoted by a "love" or "hate" between two real things denoted by "I" and "you." We have taken for granted that "The Russians are evil" and "The Americans are materialistic" refer to real facts whether true or false about real things. We have gone even further than this and taken it for granted that if we say something that sounds sensible, then it is sensible and refers to something real (whether or not it is true). For example, if we say, "The One is both immanent and transcendent!" this sounds sensible, and we might start wondering about it and get very upset because we cannot figure out precisely what "The One" is. Nor seemingly could many of the great philosophers whose works we may read to try to discover what it means. It seems to be one of the most inveterate tendencies of human thought to think in terms of things. This has been no less true in the history of science. Whenever a new phenomenon has been discovered some *thing* has been proposed to account for it: phlogiston causes fire, caloric explains heat, ether is the basis for light waves, and so on. And these names—"phlogiston," "caloric," "ether"—were taken to correspond to some real stuff in the real world, just as nowadays "fields" and "quarks" are taken to be real things.

In the 1930s a revolution in philosophy occurred, along with all the other revolutions that seem to have occurred at this time. The new philosophy proposed once and for all to place philosophy on a scientific foundation. It proposed to do this by accepting the correspondence theory of truth, and also by saying that a sentence is meaningful, makes sense, only if it can be tested—confirmed or falsified more or less directly, against sensory observation. Otherwise it is nonsense. In this way the logical positivist philosophy, as this movement is called, proposed to show that most of the "great questions" that philosophy had pondered since Plato were caused simply by misusing language. Either by creating a new ideal language, or by straightening out ordinary language, it was proposed to show that these questions would vanish. They simply could not even be asked in a language that was used properly.[2] Thus another aspect of this general program was known as linguistic philosophy.

For many decades the program of the linguistic philosophers was assumed to be possible, and language philosophy came to dominate British and American philosophy departments. It was also, for most scientists who reflected on these things at all, taken to be the foundation of science. The philosophy of science became known as "logical empiricism": "logical" because its deductive statements were to be derived from basic statements by the rules of propositional logic; empirical because the only basic statements allowed, to which the rules of logic were to be applied, must be derived from observation and confirmed by experiment. This presupposes the validity of the correspondence theory of truth: that is, the basic statements are taken to correspond to an element of "reality," a fact.

The work of these philosophers over the decades seems to have been of inestimable value for the Western tradition for at least two reasons. First, Western philosophy *had* in the nineteenth century become utterly disconnected from the ordinary lives of ordinary people. It had lost itself in incomprehensible and possibly meaningless verbiage. Thus, philosophy provided no lure at all by which ordinary people might begin to grow into a broader vision beyond the mundanity of their lives. Of course, this was not regarded as the purpose of philosophy, which was much grander, namely, to find what is ultimately true or at least to provide a true basis for knowledge. Thus, language philosophers demanded that philosophy begin completely afresh by grounding itself on the obvious truths of everyday life and developing only those propositions from these obvious truths which were *logically* incontrovertible. A sentence of a *logically* incontrovertible type would be "It is not both raining and not raining." Ironically, this sentence could quite easily be false in a certain context in an actual life situation. G. E. Moore, although he did not regard himself as founding any school of philosophy, was one of the initiators of the ordinary language movement. His response to a statment such as "There are no material objects" would be to hold up his two hands and say, "You are certainly wrong, for here is one hand and here is another, and so there are at least two material things." Or in response to "You do not know for certain that there are any feelings or experiences other than your own," he might reply, "On the contrary I know it is *absolutely* certain that you now see me and hear what I say and that my wife has a toothache." And so on. Thus the connection with traditional philosophy was thought to be severed; philosophical problems were seen to be knots in particular language games. It was supposed, partly in response to the behaviorist view of psychology, that if people could be reeducated to speak in a language in which such problems could not be spoken, they would simply never have questions such as "What is real?" and "Who am I?" and "Are matter and mind the same or different?"

The emphasis on examination of how we use language and our return to the ordinary commonsense use were, then, the first contributions of language philosophy. And indeed we will see later that some apparently profound questions can be resolved or at least thoroughly clarified by such analysis. The second contribution was that language philosophy took our ordinary idea of the relation between language and reality—the correspondence theory—formulated it as precisely as possible, and tried to found the philosophy of science and of everyday life on it. After half a century, it is now more or less clear, even to many language philosophers, that the correspondence theory of truth is applicable in such rare and artificial cases that it provides an utterly inadequate basis for understanding the relation between language and reality in human life as it is lived.

Our interest here is not to enter into disputes between schools of professional philosophers. It is rather to focus attention on a very deeply held belief, the source of perhaps many other beliefs. That is our belief that the words we use, and the sentences we say, correspond, to a large and significant extent, to something real outside of them. We believe that the words we use behave like and can stand in for the things they represent. The obverse of this, which does not follow from it, is even stronger and also widely believed: namely, that if we cannot verbally describe a particular experience, then we did not "really" experience it, or it did not "really" happen. Science depends on this principle for its claim to provide descriptions of "reality," and we normally take it for granted in all of our activities of daily life, in our intimate, political, economic, and religious conversations, and even in our private thinking. All popular science books and magazines assume it. There has recently been much discussion of the idea that physicists are "saying the same thing" as "perennial mystics" or "Eastern mystics" have always said. This is of interest if it is assumed that because they are saying the same words, they must therefore be talking about the same real thing. This assumption disregards the different life histories which lead physicists and others to write similar-sounding paragraphs, and is based on acceptance of the correspondence theory of truth. These similarities may be an indication of deeper structural correspondences, but the question becomes how to go beyond the similarities of language.

Involved in the correspondence theory of truth are three elements: the nature and structure of language; the nature of reality; and the relation between the two if they are different. A gradual disillusionment with this understanding of truth came on all three fronts. First, it was gradually admitted even by the ideal language philosophers themselves that the kinds of languages that could be constructed based on the principles of logic and direct observation were so artificial and poverty-stricken that they bore no relation to a language that any human would use. Second, the kinds of real situations that could be described if one tried to confine oneself to talking about observable commonsense elements of reality were also so trivial as to be barely human. Iris Murdoch says of Gilbert Ryle, a prominent ordinary language philosopher, "Ryle's world is one in which people play cricket, cook cakes, make simple decisions, remember their childhood and go to the circus, not the world in which they commit sins, fall in love, say prayers or join the Communist Party."[3] And according to H. Brown, "One of the striking aspects of logical empiricist philosophy is lack of detailed analysis of actual scientific theories or examples of scientific research. Rather we find analysis of propositional forms, artificial languages and occasional illustrations by reference to simple empirical generalizations such as 'all ravens are black.'"[4] The same thing applies to the attempt to apply the correspondence theory of truth to ordinary language in ordinary life—one can speak only of banalities. If we lived in a world in which we could speak only in such terms it would not be the human world as we know it. There would be no science, no poetry, and probably no humor. And one must certainly suspect the whole enterprise as having little relevance to human life, when even one of the most stalwart defenders of the correspondence theory, Karl Popper, could write of his meeting with Tarski, another great logician, "No words can describe what I learned from all this, and no words can express my gratitude for it."[5]

It turned out to be impossible to carry out this program even in physics, the most likely candidate for such a strict and unnatural use of language. To try to define "electron" in terms of basic observational statements together with logical deductions from them was absurdly cumbersome, and impossible besides, since it would mean redefining "electron" each time a new relevant observation was made. This in turn would mean redefining the interrelation of other observations, and the whole effort would become circular. As we shall see in Chapter 5, a revolution in the understanding of what science is developed out of the realization that such "basic statements" do not exist. All statements, no matter how much they seem like bare observation, are in fact theory laden, based on unacknowledged presuppositions. And this is one reason the correspondence theory of truth breaks down on the individual level: to check a simple language statement against "reality" involves an act of perception, and perception itself is governed by deeper unacknowledged language statements.

Furthermore, when we look at the actual life situations in which language is used by a speaker to one or more listeners, we begin to find that the meaning and truth of even the most simple sentence depend on the context in which it is uttered. First of all, they depend on a large range of phenomena beyond the actual words of the sentence: variations of pitch, loudness, duration, eye movements, head nods, facial expressions, gestures, body postures, and so on. These are known as paralinguistic phenomena. Take them away and you remove almost all meaning: "If the appropriate paralinguistic elements are omitted, the participants in a conversation get confused, nervous or angry; they may lose the drift of what they are saying and become more or less incoherent, and they may stop talking altogether."[6]

Beyond this, the meaning and truth of statements depend on the larger context: "The trees are red" may be true for a four-year-old drawing pictures; "The moon is made of green cheese" for a display of cheeses depicting the solar system in a delicatessen window; "I hate you, Daddy" momentarily when Daddy takes away the candy she was eating just before dinner; "He is doing very well today" for a man dying of cancer who opened his eyes and said hello to his wife; "Space is empty" for an astronaut making sure his space walking suit has a functioning oxygen system; "Time went very fast" for two lovers, and so on. George Lakoff has shown that the truth of a statement relative to a particular person in a particular situation depends on a process of categorization that rests not on the properties of objects themselves, but on interactional properties that make sense only relative to human functioning.[7]

To take a very simple example, there is a well-known riddle that goes like this: "Police are called to the scene of a crime. They find in a room a dead man who is lying on the floor, two overturned chairs, a table, and fifty-three bicycles. What happened?" There is one and only one correct solution to this riddle, and it depends on remembering that there is a famous brand of playing cards depicting bicycles on the backs of the cards. The reason the riddle puzzles people is that they immediately categorize the term "bicycle" in one way and do not realize other possibilities. Lakoff gives the example of someone who is asked to bring four chairs with him when he comes over for a discussion group after dinner. He brings a dining chair, a rocker, a bean chair, and a hammock, all of which work well for the discussion group. If the chairs had been requested for a formal dinner, however, the host would have been not at all pleased. Thus, the category "chairs" is not inherently fixed in terms of the property of the objects themselves, but defined in relation to the situation in which it is being used. Lakoff shows that we do this a great deal in our ordinary life and concludes, "Truth depends on categorization in the following four ways:

"1. A statement can be true only relative to some understanding of it

"2. Understanding always involves human categorization, which is a function of interactional (rather than inherent) properties and of dimensions that emerge from our experience

"3. Categories are neither fixed nor uniform. Whether a statement is true depends on whether the category employed in the statement fits, and this in turn varies with human purposes and other aspects of context

"4 The truth of a statement is always relative to the properties that are highlighted by the categories used in the statement (for example, 'Space is empty' highlights 'empty' and is true relative to lack of oxygen.)"

When we begin to take context into account we begin to enter a realm as complex as the world itself: "A complete theory of context would have to take into account all the knowledge that speakers have about the world."[8] So we begin to realize that language is not at all a distinct set of signs which are defined mainly by their relationship with each other, and which accurately mirror the real world. It is an interacting aspect of that real world itself, defined by it and altering it.

Natural Language: Structuralism and Functionalism

If, then, language is not merely a mirror of the world, how shall we begin to understand what is the relation between that aspect of our experience which is language and that aspect of our experience which is not language? By language here we mean not just the vocal use of words to communicate with others, which we have mainly been discussing up to now and which is the simplest to analyze. We also mean those deeper, more internal levels of verbalization, conceptualization, and imagery by which we carry on internal conversations, by which we describe the events of our lives to ourselves. As Alton Becker, a linguistic anthropologist says, "Many have argued (e.g. Croce, Sapir, Freud, Lacan, Ernst Becker, Emile Benveniste) that the imagination is an aspect of language, or at least 'structured like a language.'"[9]

Let us first see what has been the result of the study of natural languages in the past half century or more. Much of the work that has been done is often divided into three main categories: structuralism, functionalism, and generativism. Although of course much work has also been done that does not belong in these major schools, we will begin our discussion with these. We will focus on the issue of how words in a natural language derive their meaning, since this is the dominant issue in the relation between a language and the reality it purports to describe.[10]

The beginning of the modern study of natural languages is often put at the year 1916, which marked the publication of Saussure's *Cours de linguistique générale*. We might also place it in 1911 when Boas published his *Handbook of American Indian Languages*. Boas was, like Saussure, a structuralist, but he was more concerned with the description of various languages as an end in itself than with formulating conceptual schemes for providing a general understanding of language. Until this time the study of languages had been mainly historical. It was thought that in order to understand a present language, the current meanings of words, and their use in sentences, it was necessary to look back and see how that language had changed and developed from its earliest roots. Languages were regarded almost as living entities which had a life of their own and had evolved from primitive roots in a way similar to the evolution of species. In this regard, it is important for our understanding of the relationship between language and culture that it is now thought that there is no evidence whatsoever for the idea of "primitive" languages. Nor is there evidence for any directionality in the change of languages from their origins to the present day with regard to structural sophistication or complexity.

Saussure proposed that the inherent structure of a language at a given time should be studied, that its history was irrelevant, as it was to the speaker of a language. He suggested that any particular language is a coherent system which is as real as, although different from, a physical object. The language system was external to any particular individual speaker of the language and exerted constraints on him by providing a system of value which was maintained by social convention. Thus a language was real in the same way as a legal system and a political system were real. They were independent of individual people or interest groups, yet they had real causal effects on the behavior of individuals. He distinguished between the language system, a social fact, and the language behavior, the deepest expression of the language by an individual. The meaning of any unit (sound, word, or sentence) in such a language system derived entirely from its relationships to other units of that system. In fact, words could be said to form a network of relationships so that any one word was ultimately dependent on its relationship to all the others. In other words, structuralism attaches more importance to the relationship between entities, words for example, than to the entities themselves. He pointed out the arbitrariness of the link between the form and the meaning of a word, that is to say, the sound of "chair" has no relation to anything we could sit on. Therefore, the words themselves did not have any inherent connection with the objects they represented. This notion of arbitrariness combined with the notion of a language system as a coherent structure which defined itself internally led naturally although not inevitably to relativism: the idea that there were not necessarily any universal principles of sound or grammar common to all languages.

Saussure himself does not seem to have paid much attention to the relation between language as a social fact and the society or community of speakers of that language, other than to make the distinction between them. There is, however, a very strong statement of such a relationship, known as the Sapir-Whorf hypothesis. This was hinted at by American linguist William Sapir in 1928 and put in its strongest form by Benjamin Whorf as a result of his studies on the language and world view of Hopi Indians. It is put very well in this passage by Sapir: "Human beings do not live in the objective world alone, nor alone in the world of social activity as ordinarily understood, but are very much at the mercy of the particular language which has become the medium of expression for their society. It is quite an illusion to imagine that one adjusts to reality essentially without the use of language and that language is merely an incidental means of solving specific problems of communion or reflection. The fact of the matter is that the 'real world' is to a large extent unconsciously built up of the language habits of the group. No two languages are ever sufficiently similar to be considered as representing the same social reality. The worlds in which different societies live are distinct worlds, not merely the same world with different labels attached."[11]

In referring to the real world, Sapir here apparently means mainly the social world. Whorf attempted to strengthen the statement by showing that it is also true of perception of elements of physical reality such as color, space, and time. Whorf maintained that perception of physical time by the Hopis, for example, is completely different from our own, or even absent. Hopi, he says, is a timeless language, and therefore the Hopis have a very different concept of time passing, and would indeed have produced a very different physics had they been scientifically oriented.[12]

Such a strong statement is very difficult to confirm, and nowadays it is generally agreed to be too strong. However, we can certainly agree with Sapir's statement in regard to social fact. And as we will see later, our world description as embodied in language certainly has a profound influence on our perception of the physical world and thereby on social fact itself. Furthermore, we must bear in mind the extent to which we live almost entirely in socially defined reality and minimally in bare physical reality.

A further step in elucidating the relation between language structure and social context is taken by the functionalist school. This is also sometimes called the Prague School, since it began with a circle of scholars collected around Vilem Mathesius in Prague in the thirties. According to this school, the structure of a natural language is determined to some extent by the various functions if fulfills in actual use by speakers. They regarded language as a tool which has a job, or several jobs, to do and which is formed by these jobs, just as a hammer is formed by its task of hitting a nail. They were therefore interested in the esthetic and literary uses of language as well as the expressive and interpersonal uses. They regarded all these as being as significant as the descriptive function in determining the meanings of words. Clearly, if we take this point of view, the context of speech will affect its structure. If a sentence is uttered in order to convey information, then it will be tailored to what we want the hearer to learn, what he already knows, and the context of the conversation that has already been established between us. This will affect our use of words, the meaning we give to them, and the way the hearer interprets them. We saw a profound and widespread example of this in Lakoff's discussion of categorization.

Generativism

The school of generativism, started by Noam Chomsky in the late fifties, states what appears to be the opposing view to structuralism, at least in its extreme, relativistic form.[13] It developed partly as a reaction against American behaviorist linguistics, which attempted to place linguistics on the foundation of logical empiricism, just as language philosophy had attempted to do for philosophy, and as behaviorism attempted to do for psychology. The main features of behaviorism are its rejection of unobservable (to another) mental states and minimization of the role of innate motivation of any kind. It emphasized mechanism, overt behavior, and the part played by learning, of which the stimulus-response reflex was the prototype, in how animals (including humans) acquire their behavior. Like language philosophy, behaviorism had the positive effect of clearing the ground and reemphasizing the need to relate theory to sensory observation to some extent. On the other hand, taken, as it was for several decades, as a complete theory of human language, learning, and behavior, it was absurd and had disastrous consequences. In the case of language behavior, behaviorial linguistics maintained that every utterance should be regarded as a direct response to an immediately present environmental stimulus, or at least to a learned disposition invoked by the immediate environmental stimuli. In the case of language learning in children, behaviorists took the view that language was learned entirely by stimulus response as children heard their parents talk and name things for them.

Chomsky responded by pointing to the creativity of actual utterances. That is, what is actually said in any particular situation might be one of a vast range of possibilities bearing very little obvious relation to environmental stimuli. For example, you might be visiting a friend who offers you a cup of tea. You could respond, "Yes, lovely day, isn't it?" or "Yes, that will help," or "No, may I smoke?" or "No, have you stopped beating your wife?" and so on indefinitely and unpredictably. Chomsky maintained that such creativity is precisely what distinguishes man from machines, but that it follows certain rules. For example, in the above situation you would not be likely to answer, "No, wife you have beating stopped your?" These rules that provide innate guidelines for what we say have universal properties, the same in all languages, by virtue of the structure of the human mind.

The other major argument for such innate structures comes from the observation of the extraordinary facility with which almost all children learn languages. It is extremely unlikely that this could happen merely on a stimulus-response basis without any innate capability. The fact that a child of, for example, English parents, if brought up in Japan, seems to be able to pick up Japanese as easily as a native child implies that these innate structures are universal. Chomsky also suggested that these innate structures correspond to (or are identified with) genetically inherited structures in the human brain. Thus, the main task of the generative school of linguistics has been the search for such universal elements supposed to occur in the grammar of all languages.

Clearly, such a search for universals is in direct contrast to the structuralist view that universals either do not exist or are of minor importance. Generativism provides a way of understanding how language might be biologically based in spite of its diversity and independent reality. Of course, it is possible that environmental pressures caused by the social influence of language described by Sapir, Whorf, and the functionalists could bring about the genetic predisposition proposed by Chomsky. That is, even if we adopt the crude "natural selection" view of evolution, which in a later chapter we will examine more closely, we can see that the social constraints caused by language would provide a selective advantage for individuals who had begun to have some innate language capability. We will also see in a later chapter that there is a biological basis for language ability in the human cortex.

Now we can begin to see how these three major schools of thought provide us with a broad view of the relation between language, society, and the individual. To summarize: a particular language, such as French or Hopi, is regarded as having a reality independent of individual speakers of that language. This reality is not, at least at first sight, the *same* as the reality of tables and trees, but since language can causally affect these things, it must be regarded as *equally real*. Certainly the language system must be regarded as at least as real as an electron, which in fact has the major part of its existence within the language system. We might note also that within any language system, such as English, there are subsystems which to a greater or lesser extent have their own internal coherence: for example, science, economics, politics, etc. We could subdivide these even further. The way language causally affects and is affected by the physical world is through its effect on social context, conventionally maintained social values, and conventionally agreed on modes of perception as described by the functionalists and the Sapir-Whorf hypothesis and behaviorists in their best moments. Finally, over many generations language pressure may have brought about innate, genetically inherited predispositions in the organism.

Language as Metaphor

Karl Popper has proposed a similar view of the relation between physical reality, individual inner experience, and scientific discourse, although also different in some respects.[14] He calls these three elements the three worlds: "By 'World 1' I mean what is usually called the world of physics: of rocks and trees and physical fields of force. I also mean to include here the worlds of chemistry and biology. By

'World 2' I mean the psychological world. It is studied by students of the human mind, but also of the minds of animals. It is the world of feelings of fear and of hope, of dispositions to act, and of all kinds of subjective experiences, including subconscious and unconscious experience . . . . By 'World 3' I mean the world of products of the human mind. Although I include works of art in World 3 and also ethical values and social institutions (and thus, one might say, societies) I shall confine myself largely to the world of scientific libraries, to books, to scientific problems and to theories, including mistaken theories."[15]

He goes on to point out that a particular book or a particular performance of Schubert's *Unfinished* Symphony belongs to World 1 and World 3. The physical book is a World 1 object; one could use it as a paperweight or a projectile just like a rock. The *contents* of the book, which are the same for every book of that title, belong to World 3. He argues that World 2 and World 3 objects should be considered real in exactly the same way as those of World 1, that is, "if you kick them, they will kick back." The point that Popper seems to overlook is the tremendous extent to which what he calls World 1 and World 2 objects have a very large World 3 component. The bare physicality of a chair is almost negligible in relation to our—often unconscious—association with it, or categorization of it, to use Lakoff's term. Likewise, a feeling of hope or fear depends to a very large extent on our labeling it as such. The very title of Lakoff's book, *Metaphors We Live By*, concisely and poetically summarizes this viewpoint. Lakoff and co-author Mark Johnson demonstrate in example after example that even what we usually take to be quite direct statements are metaphorical. They propose that human thought processes are largely metaphorical. Their thesis is supported by so many detailed examples that it is not possible to do justice to it here. Nevertheless, here are a few:

    1. *Argument is war:* Your claims are *indefensible*. He *attacked* every weak point in my argument. I've never *won* an argument with him. You disagree, Okay *shoot!*
    2. *Spatialization:* Happy is *up*, sad is *down*. Consciousness is *up*, unconsciousness is *down*. Health is *up*, sickness is *down*. Rational is *up*, emotional is *down*.
    3. *Theories are buildings:* The theory needs *more support*.
    4. *Ideas are food:* There's a theory you can really *sink your teeth into*.[16]

The authors demonstrate how metaphors emerge from our

experience of our bodies as such and in interaction with the physical and social environments. They show that metaphors interact to form coherent structures or gestalts and that these gestalts are the cultural presuppositions in which all of our experiences are embedded. Understanding takes place in terms of entire domains of experience, not in terms of isolated concepts. By domain of experience they mean a structured whole within our experience which is conceptualized as an experiential gestalt. It is by means of conceptualizing our experiences that we pick out their "important" aspects, categorize them, understand them, remember them, and bring them to bear on future experiences as part of the background gestalt.

In summary, we could say that our extended version of Popper's World 3, which includes a very large part of World 1 and of World 2, is formed by interacting webs of metaphor gestalts. Furthermore, it is in World 3 that a very large part—perhaps almost all—of our day-to-day experience is experienced. This includes our self-interpretation as "I."

Let us turn back now to the notion of the three natures presented in Chapter 2: 1. unconditioned or absolute nature; 2. conditioned or relative nature; and 3. imaginary, conceptual or belief nature. What we are calling here World 3 (extended) corresponds to imaginary, conceptual, belief nature. Conditioned, relative nature is World 1 and World 2 stripped naked of all their World 3 elements. To penetrate beyond World 3, to experience conditioned, relative nature directly, is not easy although it is simple. It is to separate, even for a moment, from one's conceptualizations of one's self and its experiences: to see the face of a friend as it is; to hear the sound of the wind howling as it is, direct and unstained by our World 3 ideas.

This is not in any way to depreciate language and its role in our world. But when we understand its nature as being metaphoric rather than merely that of a descriptive tool, language becomes a poetic expression of the fullness of human life. This difference between language as description and language as poetry is beautifully expressed in this passage concerning traditional Japanese poetry.

"On an autumn evening there is no color in the sky nor any sound, yet although we cannot give any definite reason for it we are moved to tears. The average person lacking in sensibility — he admires only the cherry blossoms and the scarlet autumn leaves he can see before his eyes. Or again, it is like the situation of a beautiful woman who, although she has cause for resentment does not give vent to her feeling in words, but is only faintly discerned—at night, perhaps—to be in a profoundly distressed condition. The effect of such a discovery is far more painful and pathetic than if she had exhausted her vocabulary with jealous accusations or made a point of wringing out her tear-drenched sleeves to one's face."[17]

Heidegger has shown that it is our knowledge of what is present (Greek: *eon*) and our ability to *describe what is* in language, which have further and further obscured *eon* from us. Inauthentic existence is to become caught up in the descriptions, taking them to be *what is*. In his later writing, Heidegger leads the reader back through language to realize himself as an open question. Over the generations the intensity of the question, What is *eon*? has been obscured by knowledge, by our gradual discovering and explaining of *what is*. Therefore, man must now work back to the point of view from which *eon* can be glimpsed. This working back brings us to "waiting"—not "waiting for" something, but waiting so that *what is* can come to us unspoken. We could almost say that Heidegger has brought "thinking" back into the realm of contemplation. "Our sole question is, what is it that calls on us to think. How else shall we hear that which calls, which speaks in thinking, and perhaps speaks in such a way that its own deepest core is left unspoken?" "Thinking is only thinking when it *recalls* in thought the *eon* (what is present), that which this word indicates properly and truly, that is unspoken, tacitly."[18] When language and thought are in this way freed from their bondage to description, they point beyond themselves to *what is*. This is poetry. And poetry is, therefore, the highest, most human use of language.

In this chapter we have reviewed theories of the relationship between language and reality, ranging from the naive correspondence theory, that language is a precise mirror of reality, to a view of language as metaphor, poetry, creator of World 3 universes. Later we will try to work through some of our conceptualizations in relation to the physical and biological world by juxtaposing some of the commonly held beliefs about this world with some of the recent discoveries of science. It would be a pity, though, if the reader were to take these recent discoveries as something further to believe in and therefore as one more strand in the web of World 3. All we can do in a book is to talk in the metaphors of World 3. But at least we can try to let these metaphors point beyond themselves, to leave some loose ends for the reader to pull on.

Bearing in mind the power of language to distract us from the unconditioned nature of goodness, and conversely its power of pointing beyond those belief systems to the unconditioned nature, the next step is to see how some particular belief systems have arisen and decayed. This we will do, looking specifically at our beliefs about man's place in nature, in the next chapter.

4   Growth and Decay of Beliefs in Physics

"I Believe In . . ."

We are accustomed to think of the physical world in which we have our existence as an empty container in which there are things. We think that all these things exist in their own right. For example, if everything else in the entire universe were to be eliminated, the chair you are sitting on could conceivably continue to exist, a chair adrift in nothingness. Everything is fundamentally isolated from everything else, and relationships between things are only secondary to their existence. We think that, naturally since it is utterly empty, space is completely uniform, the same everywhere. It does not act on us or affect us in any way. We exist, along with a lot of other things, in the midst of nothing—void, black, cold, dead, empty. In this view, our relationship with the other objects, people, or things that fill this void is dominated by owning them or being owned by them, controlling them or being controlled by them. When AT&T says, "Reach out and touch someone," it is so poignant. Very often we think of ourselves as an uninvolved witness of this world. An image enjoyed by physicists is "The Particle Play": all the world is a stage and all the men and women merely watchers. The play proceeds according to fixed laws which have nothing to do with human presence or awareness.

Furthermore, we think that, as our life changes, it does so relative to a background, time. This time is continuous and uniform. There is no difference whatsoever between one time and another time; time has no quality. There are no breaks in time, no reversals in time. Time just flows on without us.

This is the way we conceptualize our life and organize it. This is even, at the level of self-consciousness, how we organize our perceptions, with ourselves as witness. Such a life does not have qualities; it is simply full or empty. If we live a full life we feel satisfied. But if we feel our life is not full enough, that it is becoming empty, then we feel restless and bored, perhaps even afraid that we are wasting it. So such a life can be merely empty or full, just as the space of it can have more or fewer things in it, and we can fill up its time more or less usefully.

We think that this is the objective truth. But there is another aspect of our existence. It is what we could call the feeling level, although this does not mean feeling in the sense of emotionality or gross physical sensation. (It is probable, however, that it is taking place at a fine level of sensory perception.) Rather, we mean the feeling of quality, of value, of meaning. At this level of experience, space is not uniform, or empty. It does have qualities. Space can be peaceful, wild, energetic, warm, or cold, and we can feel this. When we talk with another person, we can feel the space between us and the quality of it. When we are with a group of people and someone leaves or a new person arrives, the quality of the space changes. Perhaps we can remember back to early childhood when we felt this much more strongly: this end of the road is threatening, these woods are friendly, every corner of the garden has its own special quality.

Likewise, time is nonuniform. It has qualities which we feel. The quality of early morning is different from the quality of noon or dusk. Certain days have qualities, sometimes angry, sometimes luscious or peaceful, sometimes very rich, sometimes very speedy. The seasons have their own quality. Sometimes we can detect a change in the quality of a fleeting moment. We may even have a sense of discontinuity in time.

We have learnt to think that all of these things are subjective and therefore meaningless and nonsensical. They are unscientific; they do not tell us anything about the actual state of affairs in the world. They are all in our mind, wherever or whatever that might be. I would like to suggest that neither physics nor biology gives us solid ground for believing that our life "ultimately" takes place against a background of empty, uniform space and continuous, uniform time, no solid ground for believing that we or anything else is "ultimately" made of little, impenetrable hard things drifting in this space and time, forever separated, isolated, and obeying laws which do not spring from their nature. And science also gives us no solid ground for believing that the mind is like a mirror which is forever separated from the world, witnessing it like a spectator at the movies.

If we wish to believe something or some authority, we may discount our immediate feeling of quality in the world. But we still cannot claim physics or biology as authority. In fact, it is precisely taking science as authority beyond its domain of relevance that has led to such lopsidedness in our culture. But we cannot blame science, only our own need to believe. And this is as true, for example, of the "new physics" as of the "classical physics."

Origins: The Discovery of Nature

Now let us briefly review how it came to be that our culture has these beliefs as its underpinnings. The founders of the classical view, Bacon, Galileo, Bruno, Descartes, Newton, Locke, to name just a few of the famous ones, were at first reacting *against* something. They themselves were following the seemingly inevitable course of centuries of debate, proclamation, and condemnation. We are accustomed to believe that our modern culture, and in particular science, had its origin in that extraordinary flowering of metaphysical speculation, celebration, and careful study of nature that occurred among the citizens of Athens in the fourth century B.C. This has a certain element of truth in it, in that a very few of the writings of Aristotle were translated into Latin and preserved in a Christian context after the failure of the Roman Empire in the fourth century A.D. Later, in the eleventh century, the writings of the Greek naturalists and philosophers were rediscovered, as they had been preserved and elaborated by Islamic scholars. But for five hundred years, during a period of decline, barbarism, and despair, the thread of civilization in Europe was preserved almost entirely in the Christian monasteries. The dominant tone throughout this period, inherited from the Christian patriarchs, especially St. Augustine, was one of rejection of nature and of this world altogether. It was an utterly world-denying philosophy. This earthly realm was regarded as merely God's back garden, and parallel to it was that other world, the invisible world of heavenly vision and ideals. It was to this higher realm that all men and women, from peasants to philosophers to kings, were to direct their contemplation. As Thomas Goldstein says, "Medieval minds had been trained to look on this life as if with the bird's eye view of that higher region, where all ideals and values—the universals—had their home. Medieval philosophy had invested a great deal of sharp-edged thought in exploring the workings of that timeless dimension, tracing the invisible threads that were linking it to the limited world of human affairs. There was much profound wisdom, much to sustain one's inner security in that cultural vision, which modern Western civilization appears to have lost. Yet the natural world seemed merely a distant region of shadows. To the mind trained on eternal vision nature seemed only a darkling patch."[1]

Thus the profound awakening of the Middle Ages, as Goldstein clearly shows, was the rediscovery of nature as a realm which had its own reality, its own modes of functioning and its own regularities, regularities which, through careful use of his senses and his reason, man could come to know. Thus a duality was set up between the heavenly realm, knowable through faith, revelation, and rational thought, and the earthly realm, known through the senses. These two realms were forever separate, and there was no question of discovering the inherent sacredness and purity of this earthly realm. We might note that it is this duality at the root of our own culture which gives rise to the continuing misinterpretation, as "world-denying," of doctrines which are not based on the duality of this world versus the other world, such as Buddhism, Confucianism, and Taoism, as well as the contemplative writings of Meister Eckhart and Father Thomas Merton.

Medieval Harmony with Strains of Discord

The year 1050 is often regarded as marking the dawn of the High Middle Ages in Europe and the following century saw a veritable explosion of intellectual, political, economic, and artistic genius.[2] Kenneth Clarke, in *Civilization*, his beautiful document of the cultural growth of Europe, calls this period one of the very few periods of real genius in the whole of Western history.[3]

Contributing in large part to the intellectual awakening of this period and to the dawn of Western science was the rediscovery of the texts of Aristotle after the conquest of Spanish Islam. For the next two hundred years, it was debated whether natural philosophy, embodied in the works of Aristotle, could or could not be embraced by Christian theology. An increasing number of secular schools were established which began to teach Aristotelian science, medicine, and astrology as they had been interpreted by Arab commentators. Numerous condemnations and bannings of Aristotle's texts occurred from a sense of conflict between the dogmas of Christian faith, grounded in revelation, and the doctrines of Aristotelian natural science, based on reason and experience. Aristotle had maintained that human reason, combined with knowledge gained through the senses, could generate all possible knowledge, including knowledge of the existence of God. This undermined the theological view that God could be known only through revelation. In the *Summa Theologica*, Saint Thomas Aquinas provided a resolution, insisting that there was no conflict between faith and reason and demonstrating this in a truly grand synthesis. He showed that reason could demonstrate what the heart could know by faith and love, including logical proof of the existence of God.

Aquinas's world view became church doctrine, as well as the basis of much of natural science for the next three centuries, and continues to be an important element in Catholic theology today. Perhaps the strength of Aquinas's system was that it brought together in one image, without contradiction, an explanation of the natural world and the way in which man can come to know it, on the one hand, and the spiritual world and the way in which man journeys through it, on the other.

But it is worth remembering that in spite of Thomism's triumph as the conventionally agreed-on world view for so many centuries, there were many dissident and critical views throughout this time. Ockham, for example, a contemporary of Aquinas, had proposed that faith and reason were utterly *unreconcilable.* He agreed that we could have certain knowledge of our experiences, but he maintained that this could tell us nothing of what lay behind those experiences, and thus could in no way lead us to first principles or to God. Ockham's view laid the ground for pure empiricism and the beginnings of modern science as we know it. There was also the continuing criticism of Aristotle's theory of earthly motion by Jean Buridan, Nicholas of Oresme, and others. We will return to this later. These serve as illustrations of the tremendous interweaving of themes and counter-themes we find throughout the history of science. What has been merely an undercurrent for centuries, a reaction against the conventionally agreed-on view, becomes in the end itself the conventional view, only to be replaced again centuries later by a view remarkably like the one *it* replaced but in a new guise. The conflicts between such rival views have often been fought "with passion, misunderstanding and cross-purposes," [4] as they are today.

I will outline now the main features of Aristotle's physics that were incorporated into Aquinas's systematic world view.[5] The earth which is motionless is at the center of the universe, and surrounding it are ten spheres of extremely subtle, transparent substance. The sphere nearest the earth carries the moon, and the next six each carry the sun or one of the five planets. The eighth sphere carries the stars. The ninth carries nothing on it but it is known as the *primum mobile,* for it both moves itself and causes all the other spheres to move. The tenth sphere, the empyrean heaven, is the abode of God, and is at rest. The universe ends there. In this system, a God, the Prime Mover, is *required* by the physics; he is not an extra mystery requiring further explanation, beyond an explanation of the world itself.

Beneath the sphere of the moon, in the earthly realm, all is unstable and restless. Matter here is composed of the four elements in various combinations: earth, air, fire, and water. The natural place of earth would be at the center, and of water next to the earth. Beyond this would, naturally, be a sphere of air and then of fire. However, the four elements are not in their natural places but are all mixed up with one another. Therefore, they are constantly striving to return to their natural places, and thus there is constant motion and turmoil in this earthly region.

Beyond the sphere of the moon, all is filled with a fifth substance, the quintessence. There is no empty space, which would be impossible in Aristotle's physics. The spheres move in the most perfect motion possible, which is circular. But in order to account for the motion of the planets which do not move across the sky in perfect circles as the stars do, Ptolemy (100-178 A.D.) proposed a system of smaller circles riding on the main circles and carrying the planets. By this complicated system of circles on circles, the observed paths of the planets could be approximately predicted. Thus the universe was divided into two parts by the sphere of the moon: the heavenly realm in which all motion was circular, and the earthly realm in which all motion was chaotic as everything strived to regain its natural place.

The Aristotle-Aquinas view of motion in the earthly realm was very much in accordance with ordinary, everyday observation: an object would continue to move only so long as a force acted on it. As soon as the force stopped, the object would immediately stop moving or, if it were a projectile, would drop straight down to earth. The speed of an object would be constant if a constant force were applied. The speed would also increase as the resistance of the medium it was moving in decreased. This meant that in a void, where there would be no resistance, objects would move with infinite speed—which was one of the reasons that Aristotle thought a vacuum was impossible. The fact that an object needed a moving force in order for it to move at all meant that Aristotle's world was filled with spirits or intelligences causing all the various motions that we see. Or if something was self-moving, such as an animal, it would have to be endowed with a soul to cause its motion.

This then was the universe in which for three centuries people believed themselves to live. It was a full, living universe and one in which the soul's ascent to heaven could be easily visualized as a journey through the nine spheres to meet God. I have barely done justice in this short summary to the subtlety and magnificence of this universe and to the fine complexities of logic supporting it. Dante's epic *Divine Comedy* is a beautiful and elaborate description of it.

Along with Aristotle's cosmology, Europe had inherited from the

Arabs sophisticated systems of astrology, alchemy, and magic. These interests were reinforced in the fifteenth century by the rediscovery of the hermetic tradition—a system of natural magic attributed to Hermes Trismegistus, then thought to have been an Egyptian who lived around 2000 B.C.—and the unearthing of cabalistic, Pythagorean, and other systems of transformative practices.[6] These practices involved the study of sympathetic relationship between the various hierarchical levels of the cosmos: for example, between the solar system, parts of the human body, and various metals or plants. This is the source of the well-known motto "As above, so below." It was thought that by contemplation of various relationships in the natural world one could come to a kind of gnostic understanding or direct intuition of relationship on other hierarchical levels. There were contemplative practices involved in these traditions, particularly connected with visualization and the art of memory and with alchemy, and it is possible that the best of them did have some transformative value. It is thought that Giordano Bruno, who is usually said to have been burnt at the stake for declaring the infinity of the universe, was probably in fact so burnt because of his association with a magical order. At the same time, these beliefs were at this time by no means merely a secret undercurrent. One of the leading magi, Pico della Mirandola, was declared by the Pope to be free from taint of heresy, and in the late fifteenth century astrology and magic were orthodox, Christian concerns.

Morris Berman in *The Reenchantment of the World*, argues convincingly that the transition of the Scientific Revolution, fully embodied in the personality of Isaac Newton, was the onset of a kind of madness, of mass psychosis.[7] This was due, he suggests, to the loss of the medieval "participating consciousness" which "was in effect dedicated to the notion that real knowledge occurred only via the union of subject and object, in a psychic-emotional identification with images rather than a purely intellectual examination of concepts." Berman calls this loss of participating consciousness the disenchantment of the world, arguing brilliantly that the only way to return to sanity in our modern age is to find a genuine balance between the rational intellect and a new understanding of participating consciousness. The essence of this consciousness is, he maintains, the recognition of resemblances and correspondences between all things: that all things have relationships of sympathy and antipathy to each other. As he says, "The world duplicates and reflects itself in an endless network of similarity and dissimilarity." This world view which recognized no distinction between mental and material events is the meaning of alchemy. And we can only understand what actually took place in an alchemical laboratory as we recognize participating consciousness. He concludes, "It is not merely the case that men conceived of matter as possessing mind in those days, but rather that in those days, matter *did* possess mind, actually did so." Thus, the medieval world was enchanted, it was *actually* magical. Men and women perceived in their world phenomena which we simply no longer know.

Berman argues further that the disenchantment of the world was connected with the repression of the body of the human as organism, at the onset of the mechanical age. It was connected with the loss of the tacit dimension of immediate, felt knowledge. From the time of Descartes onwards, the role of the body, and hence of the unconscious, in perception was ignored. Thus perception itself and the action that follows from it became unbalanced, and in the end grotesque, in its distortion of man's relationships within nature. We will discuss in later chapters, especially in the work of Gregory Bateson and Alfred North Whitehead, the movement, in science, toward this participating, non-dualistic consciousness, which I have called intuitive insight.

Galileo's Antirationalism

In 1543 a quiet monk, Copernicus, published a book arguing that the earth rotates daily and moves around the sun. Although his astronomical system did not account a great deal better than Ptolemy's for the motion of the planets, it was mathematically simpler. Copernicus had made his proposal because he was disturbed by the awkwardness and complexity of Ptolemy's system and by the arguments among mathematicians concerning it. He did not in any way wish to oppose the church. However, adoption of his system did far more than simplify the mathematics of calculation of planetary positions. It swept away the entire Aquinian cosmology, as well as alchemy and magic.

It was Galileo's demonstration of the falsity of Aristotle's theory of earthly motion which enabled him to cast sufficient doubt on the whole system to remove the earth from the center of the universe. In his famous experiment in which he rolled balls down an inclined plane, Galileo showed that under a constant force (the pull of the earth) balls did not move with constant speed. In fact they increased speed (accelerated) at a constant rate. He also showed that if a ball were moving on a smooth, horizontal table top, with nothing pushing it, it continued at a more or less constant speed, slowed only by air resistance. That is, contrary to Aristotle's belief, it did not stop immediately when the force ceased.

Thus Galileo had finally, incontrovertibly, shown the error of Aristotle's version of motion—a version that had been both believed *and* disputed for centuries. The way was therefore opened for accepting Copernicus's earth-centered picture of planetary motion. Galileo was a pugnacious, charismatic, and courageous character. He resorted to every means to convince the official church and his secular colleagues to adopt Copernicus's ideas. He even gave public lectures in colloquial Italian rather than Latin, which of course made him a further irritant to those whose responsibility it was to preserve the order of society by maintaining the commonly adhered-to belief system.

Nevertheless, Galileo triumphed. The triumph of his experimental method meant the triumph of his assumptions. Thus, one of the fundamental self-contradictions of science was introduced. For Galileo argued, "Look and see what is there. Do not just argue what *should* be there based on what the accepted texts say." Yet it was Aristotle's ideas that had been based on simple observation of everyday life; what Galileo was "looking and seeing" were idealized situations: frictionless balls moving in a vacuum, for example. For Galileo only things which could be measured were to be counted as valid grounds for argument, and the language for describing them was to be the pure language of mathematics, free from human wish and whim. This became the basis for the later distinction between primary, measurable qualities and secondary qualities such as blueness and noisiness, directly given to the senses. Thus Galileo brought into science the theological suspicion of the human capacity to know things directly as they are—fundamentally the inadequacy of man to grasp complete truth which for theological doctrine only God could know. This view also introduced the sense of split between pure observing mind and observed nature.

Science today takes these contradictory assumptions as unquestionable givens. It does not realize that its own distrust of a human being's ability to know nature directly, as it is, derives indirectly from theological dogma. This is science's version of original sin. Because of this distrust, a barrier, the ideal of objectivity, is placed between man and nature. The reaction of Galileo and many others to the conventional view that true knowledge could come only through revelation, or textual authority, was absolutely necessary. It was the only healthy way for men to break out into a broader view of their world, in that particular era. However, Galileo's science, just like Aristotle's, carried the seed of its own demise, which is now coming to fruition. This seed is, of course, its fundamental distrust in the inherent goodness of human nature. The distrust of the possibility for all men and women to know the nature of their world directly is probably a far greater burden of inheritance for modern man than is the mechanistic view of the universe.

One other biblical theme incorporated into the "new" science of the sixteenth century which should be mentioned is that of power over nature. Francis Bacon proclaimed that mankind should study science in order to claim his birthright of dominion over nature and use of nature for his own glory.

Thus we inherited the image of man witnessing, controlling, and manipulating nature, but ever separate from it, never able to know it directly because of a fundamental fault in his own nature. This gives rise to such statements as this, in an otherwise excellent book on new physics: "Feynman sees the world of molecules as a potential building site for all sorts of new structures where we could build tiny devices that would perform specific tasks . . . . Molecular 'societies' could be built for human ends. The microworld is a world as vast as outer space and human mastery of that world is just beginning. Conceivably the survival of our civilization could depend on our ability to master that microworld."[8]

One might add that the threatened extinction of our civilization has conceivably also come about in part from just this attitude of human mastery. It is the imbalance in this statement that gives it its rather grotesque character, not that there is anything particularly wrong in finding out about the molecular realm, and making use of those discoveries in a harmonious way for the good of all species, including man. However, the idea inherited from Bacon that we can ultimately master nature does imply that we are separate from it and should get on top of it before it gets on top of us. This view also contains the seed of its own demise: the wish to be separate from nature, if it were accomplished, would mean extinction.

None of this detracts from the positive achievement of Galileo and Bacon and many others in sweeping away the dependence on received beliefs and suggesting: Let us look and see. Let us derive our knowledge not only from what we *think ought* to be happening based on applying principles of logic to first principles known by revelation or intuition, but also, and more important, from patterns we can detect (inductively) in what we *see* is happening. The former view is known as rationalism and the latter as empiricism.

A Non-living World

Once Galileo with his new methods had convinced natural philosophers that the Aristotle-Aquinas explanation of earthly and heavenly motion was erroneous, a new explanation was needed, since men so need explanations. This was the achievement of Newton. He realized that, if we agree to Galileo's principle that an object will continue to move in a straight line unless a force acts on it, then there must be a force constantly acting on the moon. Otherwise it would long ago have flown off in a straight line instead of continuing to move around the earth in a circle. He realized that this could be the same force as that which pulls an apple or any other object to the earth. He then calculated that the motion of the other planets would be very exactly predicted if we assume that they too are kept in orbit around the sun by such a force of attraction. This force he called gravity (meaning heaviness). The term gravity is related to the term grave, which is both an adjective meaning very solemn ("That was a grave mistake"), and a noun meaning a hole in the ground. This may account for the solemnity of the way the law of gravity was regarded by Newton and his successors. It was a very grave law.

In fact it was the grave of all of Aristotle's spheres and intelligences necessary to move the planets. Now the planets just went on moving by themselves, held in orbit by gravity. Space had to be empty because the planets appeared to experience no friction: over all the millennia of observations there was no evidence that they were slowing down. Furthermore, in order for the motion of all the planets to be included in one mathematical system, they had to be moving relative to something beyond them. Newton chose to assume that his now-empty space provided an absolute, unmoving background to which all motion could be referred. He also chose to assume that the time element in his equation was the same for all planets and thus, by extension, for the whole universe. These were Newton's *assumptions;* other assumptions would have been possible.

Again, in order to explain both Galileo's observation of moving balls and also the motion of the planets, Newton had to introduce the idea of the measure of the amount of stuff or matter in a particular object. This he called mass. How rapidly something speeds up (accelerates) when it is pushed depends on how much stuff is there, i.e., on its mass. So a minicar with a 1.8-liter engine can accelerate much faster than would a school bus with the same engine. If the bus were four times the mass of the minicar, then it would accelerate four times less rapidly. This mass should not change if we moved the object to different places, nor should it be different at different times. Therefore, space and time could in no way affect mass. Thus space and time came to form the passive, uniform backdrop to the motion of lumps of matter moving automatically according to the law of gravity.

Out of the picture, along with Aristotle's spheres and intelligences, were also all the principles of sympathetic magic, principles of correspondences between body and nature, and so on. For the law of gravitation was the only law of the universe and could now explain everything, and anything that could not be explained by it was now out. Finally, out of the picture went the prime mover and the determinate God as the lure for perfection in nature. God became the clockwork maker.

From being a living organism, full, complete, finite, and ultimately determinate, the universe became a dead nothing, with lumps of dead matter in it. It was predictable and determinate not just ultimately but by moment. Yet at the same time it was infinite.

The mechanistic view, combined with man's separation from nature, was popularized and became the basis for new political and social theories as well as theories of human mind, knowledge, and behavior. Combined with the experimental method, it gave tremendous impetus to biology and chemistry. Alchemy developed into chemistry with the introduction of the quantitative approach and loss of trust in natural magic and in the transformative power of natural correspondences. With the emphasis on isolated lumps of matter as the ultimate explanation of everything, it was inevitable that the atomic theory of the elements would be developed. This was done in the mid-eighteenth century, principally by Priestley and Lavoisier. All of the known substances of nature were thought to be made up of mixtures of pure elements. Each of these elements consisted, ultimately, of one kind of indivisible particle—the atom. The atoms were conceived of as being indivisible little balls with different kinds or numbers of hooks on them by which they could join together to make molecules of all the other substances. In 1871, Mendeleev arranged all the known atoms together in a pleasing and symmetrical table known as the periodic table.

Things were not quite as neat as had been hoped, since electricity and magnetism, phenomena which had been known all along, did not quite fit in. There seemed to be two electric forces, a repulsive force between two positive or negative charges ("like charges repel") and an attractive force between opposite charges. And similarly there seemed to be two kinds of magnetic forces, attractive and repulsive. These forces had to be added to the force of gravity, and the picture became a little more complicated. It was very difficult to explain electric currents and magnetic fields on the basis of little particles floating in empty space obeying a simple law of attraction or repulsion.

However, all seemed to be cleared up in the nineteenth century. Young and Fresnel, early in the century, had shown rather conclusively that light was more likely to be some kind of wave motion than the motion of particles. When two beams of light were made to overlap on a screen, they showed interference patterns (light and dark bands) like waves on water, which particles could not have done. Dark bands where the beams overlap could appear only because the beams somehow cancel each other out there, just as when the trough of one wave meets the peak of another they cancel. On the other hand, two particles can only reinforce each other. Thus, the appearance of dark bands between the light bands is indication of wave motion. Faraday had shown that electricity and magnetism were somehow connected: when he moved a magnet near a wire with a meter attached, current flowed through it. Finally, in a grand synthesis, almost as grand as that of Newton himself, James Clerk-Maxwell showed that light was the wave motion of a combination of electric and magnetic forces—electromagnetism.

Faraday and Maxwell's idea of fields or waves spreading through all of space fundamentally changed Newton's simple cosmology of space and matter. But the clockwork metaphysical view was left unaltered. In fact, curiously, it seemed to be confirmed in people's minds. The medium through which electromagnetic waves propagated was thought of as some kind of very subtle substance filling "empty" space. Maxwell even tried to invent a mechanical model for how light might move through it. The metaphysical assumptions of empty space with matter behaving mechanistically *in* it were by now so deeply ingrained that they were *directing* the theory making and observations of physicists.

By the end of the nineteenth century, one of the great physicists of the time, Lord Kelvin, said that physics was a constant assembly of facts which were in basic agreement with each other and of which the most important had been discovered. Another famous physicist at the time was advising his students not to pursue a career in physics since the only work to be done was adding a few decimal points to results already obtained. Implicit in these statements was of course the view that physics had, indeed, provided the final answer to the ultimate nature of the universe, and that all other sciences should be built accordingly. Lord Kelvin saw only "two small dark clouds" on the horizon. These small clouds were some minor experimental discrepancies, but they eventually gave rise to the theories of relativity and quantum mechanics.

I have given an obviously very simplified account of the growth and decay of the medieval explanation of nature and likewise of the growth of the classical picture. As I have tried to indicate at points, the actual change of such all-pervasive world explanations is a complex matter involving complete shifts of perspective of the entire community, redefinition of concepts, and subtle shifts in the predominance given to certain themes in the web. I hope, however, that the reader has picked up some highlights of the full, living quality of the medieval universe; of how the classical view was a healthy reaction to the dogmatism and rationalistic indulgence of the medieval scholars; of what were the new themes in the classical view; and of what was in actuality carried over from the medieval view. I hope, too, that you have some sense of how what was fresh in 1100 had become dogmatic belief by the time of Copernicus, and what was fresh with Galileo had become dogmatic belief by the beginning of the twentieth century, and remains so to this day. However, modern science involves more than mechanistic versus relativistic or quantum models. It involves also beliefs that science can tell us the truth about reality and that our own senses cannot. And these beliefs are inherited not from the classical view but from the medieval theological view and even before that from Plato.

The realization of how scientific belief systems change gave rise, in the 1960s, to the most profound scientific revolution in the three hundred years of modern science, namely, a revolution in our understanding of the process of scientific discovery and change. This, then, is our next topic.

5 Scientific Revolutions: Belief Systems Change

"I Believe" Revisited

When little Johnny was two years old he was driving at night with his parents. The moon was full, and the telephone wires appeared to move across it from bottom to top. Little Johnny said, "Look, moon falling." When they arrived home he said, "Look, moon come, too." When Johnny was four he no longer said the moon followed him, but he thought that the sun and the moon went across the sky each day.

When little Johnny came to be five-and-a-half he started school. He was very proud. The first thing his teacher told him was the same thing his Mummy and Daddy kept telling him, "Johnny, it is very important always to tell the truth." So Johnny told the truth most of the time, and because he thought his Mummy and Daddy and his teachers were wonderful, he thought they *always* told the truth.

Now Johnny is thirteen years old. Perhaps he could tell us what he learnt in school this year. "I learnt that scientists are finding out the truth about the universe, which is facts. Anything which isn't a fact that the scientists have found out is just an opinion and we don't have to believe it. But we have to believe facts, because they are true. They find things out by the scientific method, which has four steps:

"1. Make unprejudiced observations and discover facts
"2. Form a theory or hypothesis to explain the facts
"3. Make a prediction from this theory
"4. Test the prediction by making another unprejudiced observation."

"And what else did you learn, Johnny?"
"I learnt about space and time. I learnt that space is cold and black

and empty and dead and goes on and on and on. And I learnt that time is the same everywhere. It flows from past to present to future. It has no beginning or end and everything moves along in it."

"And what else did you learn, Johnny?"

"I learnt that things are made of matter. Matter is made of little particles that stick together to form atoms, and the atoms stick together to form bigger particles. And there is lots of dead empty space between the particles. And the particles obey laws. The laws never change and the particles always obey them. So the world is like a clockwork. A long time ago the big particles got very complex and somehow became alive. And the living things began to compete with each other and the strongest ones survived because they could get the food and the others couldn't so they died. This is called 'survival of the fittest' and 'natural selection.' And to help them survive, the animals got brains, and because it got harder and harder to survive, more and more clever ones were born, and they killed the others. And then they got intelligent and that is how humans happened.

"And I learnt how we see. There are these waves which are made of electricity and magnetism. And for every color, the waves vibrate at a different frequency. When we see the color red, that is because waves at a particular frequency are bouncing off something and going into our eye. Our eye has a retina like the film in a camera, and each point on the retina is connected to a point in our brains by the optic nerve and that is how we see. It is like looking at a photograph. And it is the same with hearing and tasting and so on. Our brain is like a computer, that is how we think. Our mind is in our brain, and when we think things, we are thinking what happens in our computer brain. When we get angry, that is because we have aggressive instincts to protect ourselves and our children against our enemies. And when we love someone, that is because we have a sex drive so we can have children and perpetuate our species. And people have a religious instinct because they are afraid of what they don't know. And you only help someone else when it is good for you. And anything you don't know, scientists will find out and tell you in the end."

So this is what Johnny learns in science class when he is thirteen. And as he gets older, year by year, he fills in the details in very sophisticated ways. If he continues to study physics in high school, he begins to learn about relativity and quantum physics. But he has already been convinced that science can explain the universe and that what his common sense tells him about space and time and mind and matter is confirmed by science. The point here is not to suggest that we should start teaching thirteen-year-olds quantum physics and holographic paradigms. The point is that we ask them to believe the *reality* of these things, and that we too come to *believe* in such things.

We say we know that these things are true because they work. They have practical results. But what of immediate experience? What of our experience of time, memory, perception? What we believe is shown more in how we live our lives than in the sophisticated theories we can expound. On the whole, most of us, including professional scientists, do tend to live our lives outwardly as if the world as Johnny described it was all we know and all we can ever know, in fact all that is. As physicist Paul Davies says, "These statements [of modern physics] are so stunning that most scientists lead a sort of dual life, accepting them in the laboratory, but rejecting them without thought in daily life."[1]

We live as if our bodies were isolated objects; therefore, we lose our health-giving connection with the earth. We live as if we existed in dead, empty space; therefore, all our energy and insight must come from within, and we constantly feel overcome with anxiety lest our energy run out. We live as if time did indeed flow from past to future; therefore, we do not rest in this moment at all. We live as if our minds were located somewhere in our bodies and arose from them; therefore, we fear death as terrible extinction. We live as if we were observers in a world of objects which are unchanging from moment to moment and which we perceive as a camera takes a photograph; therefore, we never really look, listen, taste, smell, touch. We live as if our bodies, emotions, and environment obeyed mechanical laws which we can only go along with or struggle against futilely, as if there were no way we could open beyond this; therefore, there is no point in training except for survival or entertainment. Our perception becomes reduced, we live as if our conditioned beliefs were the only truth, and we feel the notion of unconditioned goodness as a threat to our sanity.

Others Also Believe

At other times, in other places, people have believed different things about the world they lived in. From their point of view, *these* beliefs were true because they too worked. Benjamin Whorf was one of the first anthropologists to show the different world views that American Indians have. We are all familiar now with the many different ways of understanding the world that have been reported by anthropologists over the past few decades. Here are a few small reminders.

In medieval times, churches were built with thick stone walls, in the belief that the stones had actual power to protect the sanctity of the church. In fact, all objects had a living quality, or power of their own kind. "The Romanesque church, with its strong stone walls and its squat but massive tower was God's stronghold on earth. Heré God alone was lord. The house of God, the church, offered shelter, protection and justice to man, perpetually persecuted for his sins. Negative energies were transformed into positive power by penance, confession and the sacraments and men were led towards healing and salvation. Evil was banished beyond earshot of the church's bell. For here in the consecrated church was a source of divine power stored up in the sacramental presence of Christ the king and in the power radiating from the tombs of the church's particular saints and from any other holy relics it might possess, however infinitesimal."[2] Medieval historian F. Heer comments, "Medieval piety has often been reproached for its materialism, its reversion to magic and sorcery, its lack of genuine spirituality, its vulnerability to superstition. Such charges are understandable coming as they do from people whose experience of life has been quite different, but they do not do justice to the realism and vitality of the faith this piety supported."[3]

An anthropologist told me a story of a man in Java who was rather a reckless car driver. All his friends used to say to him: "Be careful, one day you will have an accident." One day he knocked over and seriously injured an old lady. In his chagrin his one response was: "Never again will I drive on this day." By "this day" he meant any day in his cyclical life which would have the same quality as this one, and he noted it in his diary. But to him it was *this* day. This man, of course, knew the linear aspect of time—day follows day. After all, he kept a diary. At the same time he felt, very personally, the truth of the cyclical aspect of time.

The great civilization of China certainly had a profound understanding of the natural world and of the human body, which gave rise to their highly effective system of medicine. Yet it is utterly different from our own. According to T. J. Kaptchuk, director of the Pain and Stress Relief Clinic of Lemuel Shattuck Hospital in Boston, who earned his doctorate at the Macau Institute of Chinese Medicine, "The difference between the two medicines is greater than that between their descriptive language. The actual logical structure underlying the methodology, the habitual mental operations that guide the physician's clinical insight and critical judgement differs radically in the two traditions. What Michel Foucault says about medical perception in different historical periods could apply as well to these different cultural traditions: 'Not only the names of diseases, not only the grouping of systems were not the same; but the fundamental perceptual codes that were applied to patients' bodies, the field of objects to which observation addressed itself, the surfaces and depths traversed by the doctors gaze, the whole system of orientation of his gaze also varied.'"

Kaptchuk relates that "the two different logical structures have pointed the two medicines in different directions. Western medicine is concerned with isolated disease categories . . . starts with a symptom and searches for an underlying mechanism, a precise *cause* for a specific disease. The Chinese physician in contrast directs his or her attention to the complete physiological and psychological individual. All relevant information, including the symptom as well as the patient's other characteristics is gathered and woven together until it forms what Chinese medicine calls a pattern of disharmony. The therapy then attempts to bring the configuration into balance, to restore harmony to the individual."[4]

The Kung San, a small hunter-gatherer tribe living on the edge of the Kalahari Desert, practice a healing ritual in which the whole tribe takes part two or three times a week. During these rituals various members of the tribe enter a terrifying state of mind and body of very high energy, which they are helped through by members of the tribe who have previously been through it themselves. If they are able to go through this state, they then enter a state (*kia*) in which they become, temporarily, healers. Over 50 percent of members of the tribe go through this state at some time of their lives, and it is considered to be a necessary aspect of being fully adult. When a tribe member was asked what happens during this process, Richard Katz, the inter-viewer, reported the following exchange:

"Kau Dwa is teaching me. . . as I struggle to maintain my Western notions of reality. 'Kau Dwa,' I ask, 'you have told me that in *kia* you must die. Does that mean *really* die?'

'Yes.'

'I mean *really* die.'

'Yes.'

'You mean die like when you are buried beneath the ground?' I am struggling with my words.

'Yes,' Kau Dwa replies with enthusiasm. 'Yes, just like that!'

'They are the same?'

'Yes, the same. It is death I speak of,' he affirms.

'No difference?' I almost plead.

'It is death,' he responds firmly but softly.

'The death where you never come back?' I am nearly at the end of my logical rope.

'Yes,' he replies simply, 'it is that bad. It is the death that kills us all.'

'But the healers get up, and a dead person doesn't.' My statement trails off into a question.

'That is true,' Kau Dwa replies quietly with a smile, 'healers may come alive again.'"[5] Healers say that in *kia* they "see things as they are" and become more genuinely themselves.

The Kung San way of life was the universal human way of life until 20,000 years ago and, for them, it appears to have changed little. Yet, according to the many anthropologists who have visited and observed them over the years, their world view is thoroughly complex. As Richard Lee, an ecologist who has lived with the Kung and speaks their language, reports, "observers are attracted by their extraordinary culture, their narrative skills, their dry wit and earthy humor and the rich social life they have created out of the unpromising raw materials of their simple technology and semidesert surroundings." The Kung world, like the medieval and Chinese worlds, is in every significant way as *actual* as our own. This includes their healing ritual which cannot be separated from the belief-system which interweaves it.

So all of these represent different ways of understanding and therefore of living our lives. Each seems to have provided the belief system for a genuinely human society with no less compassion and insight than ours.

We expect little Johnny to understand and believe our story as much as any society ever expected its young to understand and believe its own story about how the world began and functions. But when little Johnny grows up to be Mr. Big John, he no longer remembers believing that the moon came with him, he no longer remembers how he came to believe the earth goes around the sun; he no longer remembers how he came to believe everything is made of "matter"; he no longer remembers how he came to believe his mind is in his head. He thinks it is obvious. He lives in a cold, hard world of isolation and struggle. Big John has lost little Johnny and does not know where to find him. In losing little Johnny, he has lost his *felt* connection with his world.

It is perhaps fortunate that, it now appears, we may at a certain level not have forgotten what we believed when we were two and four years old. It has been shown, for example, that 80 percent of high school students, even after taking a course in physics, still explain the motion of a projectile (a thrown ball, for example) in a way closely similar to the way pre-Newtonian medieval philosophers explained it.

This explanation may be "wrong" in terms of classical mechanics, but there is something "right" about it in terms of our felt connection with the way things work in nature.[6]

There appear to be many, many levels of interpretation and memory involved in every act of perception. As many a psychiatrist also knows, if we were to completely separate such multiple levels of interpretation from any connection with our waking life, or to condense these levels into one and become completely literal, we would be on the road to insanity or disease. This was perhaps the primary clinical discovery of Freud: the causal effect of the "unconscious" on waking life. While Freud's specific explanations have received many refutations and revisions, this primary discovery remains unchallenged. At the same time, to be able to include such levels of interpretation is apparently a prerequisite for genuine creativity. Artists know this. But genuine scientists also know it: Kekule, Einstein, Niels Bohr are well-known examples, and the literature abounds with others.[7]

Scientific Method: "I'll See It When I Believe It."

The so-called scientific method which, Johnny learns, is how scientists work is a fabrication. No scientist who has written about how they work has described it in this way. Any scientist who tried to work like this would probably produce little of any value. As we shall see, it is extremely unlikely that there can ever be such an event as a pure, unprejudiced observation. Observations are always made, first, within the historical and cultural context which has given someone a reason for making an observation at all, and, second, within the context of the person's own belief system and his own organism which in turn affect the result of his observation.

What I am pointing to here is that somehow we may have gone off the track in believing that any particular set of descriptive statements, interpretations, or explanations about our world can be the "true" one, with the implication that all others are "false." It does not make any difference whether these statements are made in the name of science or not, or in the name of "classical" science or "new" science.

This is not to criticize the original inspiration of the scientific world view and method. It is pointing to the way some of the milestones on the scientific journey have been picked up and incorporated into the fabric of our education and upbringing as fixed beliefs. At any period in the history of science, there have been interpretations rippling under the surface of the view generally agreed upon, which question that view. For example, in the eleventh century, the texts of Aristotle were being discovered and explored, to be fully accepted and incorporated in Aquinas's doctrine two centuries later. At the same time, the theory of impulse was being elaborated from these same texts, which gave rise to Galileo's theory of motion and proved to expose the flaw in Aquinas's doctrine.

The inspiration of courageous men such as Bacon and Galileo swept away hypocrisy, superstition, and reliance on authority as the source of knowledge; at least this is what we learn at school. Galileo Galilei (1564–1642) lived at a time when the established belief was that the earth was at the center of the universe and therefore did not move, and that the sun, moon, and stars moved around the earth on perfect crystal spheres. They themselves were also perfect spheres. This medieval world view, developed to its high point by Aquinas and his colleagues in the late thirteenth century, was a beautiful and satisfying one. It brought within one image an explanation of the physical world and the spiritual world.

Galileo heard of an instrument which had been made in Holland that magnified distant objects. He constructed such an instrument and, looking through it, saw that the moon had craters. He called all his colleagues together and said: "Look through my telescope; you will see that the moon is not a perfect sphere." Some of the colleagues replied: "We do not need to look; we *know* it is perfect, and therefore it cannot have craters." Others looked but refused to believe they saw craters on the moon, saying that the telescope was faulty. Galileo was imprisoned and subjected to the Inquisition for saying that the earth moves, and that the moon and planets are not "perfect" spheres. Yet now such men as Galileo are the fathers of our own fixed beliefs. Now we are taught to believe in the law of inertia and in gravity just as those colleagues of Galileo believed in perfect spheres. Newton proposed absolute space and universal time as merely hypotheses which were necessary in order to complete the formal mathematics of his system. By the beginning of the twentieth century any proposal which questioned the absoluteness of space and time and the absolute, independent existence of real particles within this space and time, following immutable laws, was regarded as mere quackery.

Newton himself did, however, believe and proclaim that in his law of gravitation he had revealed the word of God. Contemporary and subsequent philosophers took up this mechanical view of the universe and applied it to interpret mind, education, politics, economics, ethics, art, and religion. Thus, an essentially limited principle which had been successful in explaining the motion of the planets, the tides, and balls rolling down inclined planes was taken as the final ground of everything in human life. The problem with our science, then, is not that through certain methods we have been able to discover certain cause-and-effect relationships in certain very limited and often unnaturally manipulated circumstances. The problem is that these principles are taken as ultimate principles which can explain phenomena outside of the limited region in which they were conceived, and to exclude legitimate observations which do not fit the scheme.

Recently I described to the director of a major nuclear physics institute the results of some of the research in precognition which seemed to be particularly well carried out and to give apparently unambiguous results. I was wondering how physics might adapt to incorporate such phenomena. This gentleman, a kind and friendly person, was interested in the practice of meditation because, he said, "it follows the experimental method; just look and see." His response to my query about precognition was, "There are some things we *know* are not true, and precognition is one of them. Therefore, in this case, experimental data is irrelevant." This is an instance of "I'll see it when I believe it," or perhaps, "I won't see it because I don't believe it."

This principle, "I'll see it when I believe it," appears to be a very deep principle of the organism, a principle which governs much of our perception. Thus, the actual behavior of scientists in relation to observation seems to parallel the inherent predisposition of the organism. In the understanding of the history and meaning of the scientific endeavor, this realization is bringing about a revolution which has more profound consequences than the theories of relativity and quantum mechanics together. Simply, this view proposes that the conventionally agreed-upon belief structure of any particular group of scientists forms an inseparable part of their practice of science and determines which observation will be acceptable to them and which will be rejected. From this point of view, then, science is not necessarily finding out "the truth," but is merely confirming or refuting its agreed-upon belief structures. In addition, belief systems change only when, in confrontation with reality, error is revealed. Such belief structures may be gradually altered or may change in sudden shifts only under tremendous pressure from large numbers of rejected observations. This revolution is happening as a belated response to the tremendous shock that relativity theory and quantum theory caused to the rather pretentious nineteenth-century belief that we finally knew the truth about almost everything. The first response to this shock was to try to use the correspondence theory of truth to shore up the foundations of science. As we saw in Chapter 3, this effort, which lasted for well over a quarter of a century, has finally failed.

Observations May Refute But They Cannot Confirm

Karl Popper effectively initiated the revolution when, in *Logic of Scientific Knowledge*,[8] he pointed out the fundamental differences between confirming a theory and refuting it. No matter how many observations in agreement with a theory that we make, even billions, there could always be one more which would refute it. Newton's theory of gravitation is just such an example: over the course of two hundred years, there must have been thousands of astronomical observations confirming the theory. Yet it took only one to refute it—Eddington's observation of the bending of starlight near the sun during a solar eclipse, which was in agreement with Einstein's prediction, but not with Newtonian calculations. Thus, we can never know for certain that a theory, a particular description of nature, is "true," but we can know for certain that it is false on the basis of just one observation (provided we trust the observation, a proviso which strikes to the heart of the matter, as we will see shortly).

To take another example: if you tell me, "All Australians have black hair," I could go to Australia and start checking people's nationality and the color of their hair. I may count three million Australians who have black hair, but I can still not be sure that you are right. The next Australian may be blond, and this one observation would immediately refute your theory. Of course, actual situations in science are more complicated than we have implied. Many theories may interlock in relation to an observation, so that we may not know *which* theory is refuted. Furthermore, subsequent observations, or corrections of theory, may refute the very theories on which our refuting observation was initially based. Thus, we can never even be finally sure that a theory *is* refuted. Popper's view is then this: "Thus theories are our own inventions, our own ideas, and when they clash [with reality] then we know there is a reality: something that can inform us that our ideas are mistaken. (Incidentally, this kind of information—rejection of our theories by reality—is in my view the only information we can obtain from reality: all else is of our own making. This explains why our theories are all coloured by our human point of view, but less and less distorted by it as our search goes on.)"[9]

In spite of his remarks in parenthesis, Popper did not quite stick to the simplicity of his fundamental insight that "the rejection of our theories by reality is the only information we can obtain from reality." He maintained beyond this that our theories are getting closer and closer to the truth. "Our main concern in philosophy and in science should be search for truth. We should seek the most urgent problems and try to solve them by proposing true theories or at any rate by proposing theories which come a little nearer to the truth than those of our predecessors."[10] Popper developed a theory of truth based on probability. The more times an unlikely theory has been confirmed, the higher the probability that it is true. This is basically what he meant by a theory's being "a little nearer to the truth." In spite of efforts by Popper and his disciples for many decades, there is really no satisfactory way of deciding how we could possibly assign a "probability of truth" to a theory. Nevertheless, no matter how high we place this "probability of truth," still one observation could refute it. So, if we follow Popper's earlier logic, all we can really say is that we should seek to solve the urgent problems by proposing theories that are at least a little further from error.

There is a profound difference between the two statements: that a theory is a little nearer the truth and that a theory is a little further from error. Popper is able to jump from the second to the first by holding onto the classical idea that a theory is true if and only if it corresponds with the facts or with reality (the correspondence theory of truth which we looked at in a previous chapter). By holding to it Popper stands between two worlds: the world in which there is one "reality," which consists of "fact," and the world in which rejection of our theories is all we can know about "reality."

In the sixties, several theorists, principally Toulmin, Hanson, Kuhn, and Feyerabend, took the decisive step beyond Popper in pointing out that the kinds of theories and experiments that scientists formulate depend on their presupposition—their belief system.[11] They also showed that the belief systems themselves become fixed not by some process of objective experimentation alone, but also by a shared agreement amongst that particular professional group of scientists as to what to allow and what to reject as valid observations and what are to count as facts. Thus, what "reality" is now refuting is not just our theory, but the presuppositions behind the theory and therefore the observation and experiments which themselves are used to validate or refute the theory. Consequently, we become tied into a self-perpetuating or self-refuting structure: So long as our theories are not refuted by observation, everything goes smoothly. But as soon as a significant number of experiments begin to refute the theory (and therefore its presuppositions and therefore the refuting experiments themselves), then the whole structure has to radically shift. Because of the self-reflexive nature of this process—observations refute presuppositions which directed the observations—such shifts are *internal* reorganizations of a cohesive structure, rather like an amoeba's absorbing and digesting of foreign material. These radical shifts are what Kuhn calls scientific revolutions: for example, the Copernican revolution from an earth-centered to a sun-centered universe, and the relativity and quantum revolutions.

Kuhn defined two kinds of science: normal or conventional science, and revolutionary science. In normal science, scientists are working within a particular, conventionally agreed-on belief structure, or set of presuppositions. Normal science is self-confining and tells us nothing about reality. In revolutionary science, the presuppositions conventionally agreed on are no longer adequate; observations which radically refute such presuppositions cannot any longer be ignored; and a complete shift of presuppositions takes place. At this point, normal science takes over again.

An analogy for this overall, complete switch in belief structures is the series of familiar perceptual illusions. For example, Figure 6, in Chapter 12, can be seen as the profile of an old woman with a large nose or the almost back view of a young lady's head. At first we can see only one of these interpretations, for example, the old woman. Then, in a sudden shift, perhaps brought about by seeing the nose as a chin, we see the other interpretation. Once we are able to see both, we cannot *not* see both. We will discuss in Chapter 12 how this simple illusion, and others, has given rise to a profound alteration in our understanding of perception. And Kuhn suggests that the way scientists as a group perceive physical reality changes analogously.

It is an open question whether such shifts of interconnected presuppositions occur rarely, suddenly, and rather dramatically, as Kuhn suggests, or whether they happen frequently and quietly. Probably both are the case, just as a series of small earth shifts can contribute to a final earthquake. However, there is general agreement, at least amongst those philosophers of science who do not still cling to the correspondence theory of truth, on some more or less strong version of the following three points.

1. Observation is theory laden. The belief system influences how one views, describes, or interprets the world. Therefore, what one observes depends on one's theories and expectations.
2. Meanings are theory dependent. The meanings of descriptive terms (e.g., electron, wave) used in a theory change as the

theories change. Therefore, as theories change, the "real things" which they purport to be about change their qualities as well to some extent.

    3. Facts are theory laden. What counts as a fact depends on the belief system associated with a theory and, therefore, ultimately on the group decision of the community of scientists holding this belief system. Facts are not the ultimate standard of "reality."

It is only when the old presuppositions clash with reality and the new have not yet become convention that we are learning anything about reality at all. But no positive description of this reality can be given: it is only that which refutes our theories. However, it is just because there *is* that which refutes our conjectures that we cannot resort to pure relativism: the idea that "anything goes," that if we believe hard enough in faeries there will *be* faeries, just like that. Nevertheless, other cultures, whose journey out of error has taken a different path, may have come across aspects of relative reality which our culture has no idea of. Clearly the change in our view of what science can tell us about a reality beyond itself is itself revolutionary. It is too naive to believe that scientists are making true statements about facts which are a closer and closer approximation to "the truth" which is the one final description of reality. Yet we continue to relate to our children the stories of science as if they were true. And we continue to propagate this view in popular science magazines and books, whether of the so-called leading edge or mainstream variety.

In January 1981 I met a boy, who was then 8 years old, together with his parents, who told this story. When the boy was three months old he began to speak as an adult. He described to his mother, a child psychiatrist, the intrauterine life and the birth process (which were corroborated in details by his mother's memory of his birth). At the age of two he gave his mother a list of more than twelve languages he wished to learn. After about two years he told his mother that he had solved the problem: he had wished to know whether these languages had a common root and had found that they had (the Indo-European root). At the age of six he had weekly meetings with a Nobel prize-winner in astrophysics, as a result of which the professor published a new model of the structure of the Milky Way. His father, a professor of biochemistry, said that by the time the boy was three, the boy had pushed him beyond his own capabilities. His piano, violin, and ballet teachers all, independently, told the parents that they felt the boy had maestro potential in these fields. The parents, of course, had tremendous difficulty in finding adequate tutors for their son. The parents had come to visit a Tibetan meditation teacher because the boy had also reported memories of other situations which appeared to be before his intrauterine life and in a time and place he could not have known. He also reported extremely strange images between these memories. The mother later recognized a significant likeness between these strange images and those described in the *Bardo Thodol (Tibetan Book of the Dead)*. Apart from the meditation teacher's advice, the parents seemed to want nothing, nor apparently were they trying to prove anything. They were very concerned about the boy's education.

The boy seemed to both confound and confirm theories of mind/body relationship in child development. He seemed to have been born fully mature in his ability to see, understand, and verbalize his experience, yet he nevertheless went through some of the classical developmental stages of infanthood and was able to describe these to his mother. It would seem that he would be of extraordinary interest to the professional child psychologists. However, the mother had written a report of the child's early months and submitted it to a major conference on the experience of childbirth. She received the paper back with the comment that it was unacceptable because (1) the conference was a scientific one and therefore not interested in fantasy, and (2) the mother's name was listed before the father's in the authorship. Possibly the conference organizers' thought process went, "Since we *know* such things cannot happen, therefore this did not happen. And besides, the main author is a woman." This is how most of us think: "Since I *know* the world is not like this, what I saw cannot be." We see here an example of a principle which we will come across again and again: that the old adage "I'll believe it when I see it" can also be reversed. A more accurate description of how we perceive our world would be "I'll see it when I believe it." We will see that this is the case with individual perception. It also seems to be the case on a group level: a whole group rejects as legitimate observations those which cannot be accommodated by the group's common presuppositions.

As another example of this principle, we might consider the response among scientists to reports of precognition. There is tremendous controversy in the scientific world with regard to precognition; discussion often becomes vehement and extremely ill-mannered. There seems to be a strong need on the part of conventional science to exclude such phenomena from consideration as legitimate observation. Kuhn and Feyerabend showed that it is always the case with "normal" or conventional science that observations not confirming the current belief system are ignored or dismissed. The colleagues of Galileo who refused to look through his telescope because they "knew" what the moon looked like are an example. This of course means that such observation *should* be submitted to especially careful scrutiny. And since precognition phenomena *have* been associated throughout their history with occultism and entertainment, it *is* necessary to take careful precautions against fraud or self-deception. However, it appears that this is now being done.

Another reason why it is difficult to discuss such phenomena within the realm of normal science is that the best data are by their very nature anecdotal, one of a kind. they consist of individual stories from individual lives. Now, science is concerned with general patterns and repeatability of results; therefore it is inherently unable to deal with unique events, except insofar as one such event contradicts current theory. But in actuality all moments are unique. No moment is repeated exactly in its entirety, although some patterns persist from moment to moment. This inability to deal with unique events is one of the greatest shortcomings of science as we now know it, and the point at which it loses contact with the actual world and moves into the ideal world of abstraction.

Let us then look at some of the available anecdotes relating to precognition:

1. On 21 October 1966, a coal tip slid down a mountainside in the mining town of Aberfan, Wales. It buried a school, killing 128 children and 16 adults. On the evening of 20 October, a woman reported having a waking dream which she told to six other people. "First I saw an old school nestling in a valley, then a Welsh miner, then an avalanche of coal hurtling down a mountain . . . " This dream took place two hundred miles from Aberfan. Another person, seven days before the disaster, told two friends: "I had a horrible vivid dream of a terrible disaster in a coal mining village. It was in a valley with a big building filled with children. Mountains of coal and water were rushing down the valley burying the building. The screams of the children were so vivid I screamed myself." There were at least two other, similar documented reports before the catastrophe.[12]

2. Twenty years ago, a well-known British poet and novelist, J. B. Priestley, announced on BBC-TV that he was conducting an investigation into unusual experiences in regard to time. He invited viewers to write to him and received thousands of letters. He had a team of trained, skeptical investigators whose job was to eliminate fraud and obvious error and reports which could be explained in a usual way. He was left with reports which the investigation simply

could not eliminate on these grounds, which were published in *Man and Time*.[13] Here is one of Priestley's reports, concerning Air Marshall Sir Victor Goddard:

"While flying in mist and rain over Scotland in 1934, Goddard saw what should have been Drem Airfield below him. But instead of the disused hangars among fields that Drem was at the time the airfield appeared to be in working order, with blue-overalled mechanics among four yellow aircraft. Four years later, the details of Goddard's experience were exactly fulfilled: The airport was rebuilt, training aircraft were then painted yellow (instead of silver, as formerly) and blue overalls had become standard wear for flight mechanics."

Hundreds of such well-documented anecdotes have been reported, and perhaps thousands more have gone unreported. To be sure that there is not some unwilling self-deception or wishful thinking involved, it is important that the precognitive dreams be written down or at least told to someone else before the event. And specificity of detail in the dream is also important to be sure that it is a genuine precognition of a specific event. Thus one type of well-documented precognitive event is the precognition of very dramatic, often catastrophic events, usually involving people. And another type is events which are basically insignificant and of no particular interest or benefit to the precognizing subject.

We hear such reports over and over again. Psychiatrists such as Carl Jung and Alex Comfort have pointed out that precognition is reported especially in situations of psychological counseling when people are perhaps less afraid of being considered silly, and more willing to notice and speak of their dreams and fleeting images.[14] It is the mundane character of such reports that lends credence to them. They occur by chance, we cannot manipulate them, and they have no particular importance. But how are such things possible according to what we believe about space, time, mind, and body? So many ordinary people have perceptions of time like the air marshall's, but because they have no way to fit them into their received world view they dismiss them or keep them very private for fear of being thought silly.

We should note that, especially in the case of precognition of dramatic events, the precognition is not necessarily of future actualities. It may be of a future potentiality over which we might have some control. In the case of the Aberfan disaster, even had the various precognitions been recognized as such, it would have been possible to act on them only if they had related to a specific coal mine. However, Michael Shallis, author of *On Time*[15] and tutor in physics at Oxford

University, gives an example in which a dreamer dreamt that he was driving down a particular road when a child ran out in front of the car and was killed. Shortly afterward the man found himself driving down the same road under the circumstances of the dream. The child did indeed run in front of him, but the driver had already recognized the circumstances of his dream and was applying the brakes. The child was saved.

Now let us look briefly at work that has been done under laboratory conditions to test precognition. The early work by J. B. Rhine was certainly open to criticism of sloppiness and not demonstrating adequate precaution against fraud. However, in the past two decades much more careful work has been done which, according to all the standards of careful science, demonstrates seemingly strong evidence for precognitive events. This work is reported in *Explaining the Unexplained*, co-authored by the esteemed elder of British psychological testing, Hans Eysenck (author of a well-known book, *Sense and Nonsense in Psychology*).

To give just one example of such experimental work: Helmut Schmidt, a physicist working for Boeing Research Laboratories in Seattle, has built a machine that tests precognition without the intervention of a human recorder. The machine fires a series of four lights completely at random, the order of firing of the lights being determined by the random disintegration of radioactive strontium 90. A subject is asked to guess which of the random lights will light next. The subject makes his guess by pressing one of four buttons. The guess can be mechanically compared with which light in fact lights up, and the result of a whole series of guesses can be mechanically recorded. The machine has been extensively examined for possibility of fraud or hidden bias. None has been found. Schmidt's experiments are very impressive, although it is certainly possible to find logical loopholes in them, such as the definition of "randomness" and whether it is precognition or psychokinesis that is being measured. From all these experiments, Eysenck deduces that the odds in favor of there being some kind of precognitive effect are so great that if we were to write how many millions of millions to one these odds are, the "millions" would cover several lines of print.

Clearly, precognitive events cannot be understood if we assume that the Newtonian or "commonsense" view of time—the unbroken straight line—is the only aspect of time. Many speculative theories of precognition, involving especially higher dimensions of time or quantum probabilities, have been proposed in the past fifty years. However, possibly the view of Michael Shallis is the best one for the moment. This is simply to recognize that time is multifaceted. One of its facets is its apparent linearity, and another facet is the precognitive effect of some future possibility on the present.

The point here, of course, is not to suggest that we should or should not "believe" in precognition or other so-called extrasensory perception. The point is that the irrational and sometimes fanatical attempts to deny that such observations have occurred many times are a symptom of the very powerful grip that the linear view of time has on conventional science. Sargent reports that a colleague said to him after a discussion of his work on precognition, "The results you presented would convince me of anything else, but this: I just *cannot* believe it and I don't know why."

A Role For Science Beyond Propagating Beliefs

What scientists seem to be doing is weaving a complex network out of their imaginings, conjectures, and theories, which opens beyond itself only at rare points. And what happesn at those points seems also to be influenced by their own imaginings. For a moment we stop in our imaginings and realize that we were in error. From there we take off again, but for a moment there was a glimpse of relative nature, beyond our collective thought process, imaginary nature.

When men first began to wonder about "reality," they already had a tremendous wealth of stories, explanations, presuppositions, and organic predispositions. The journey of knowledge is a journey of bringing these to the light of discourse and revealing falsity so that we may see more clearly beyond them. "All acquired knowledge, all learning, consists of modification of some form of knowledge or disposition which was there previously, and in the last instance inborn dispositions."[16]

I have pointed to the fundamental misunderstanding that has been at the basis of science for three centuries, namely, the misunderstanding in the relation between scientific description and "truth." This is not in any way to detract from the beauty and grandeur and wonder of science. On the contrary, it is to restore such beauty, grandeur, and wonder, but in a larger context not based on this fundamental misunderstanding.

How then might we view the contribution science can make to the journey of our life? We might perhaps look at this problem in terms of the three natures of previous chapters, namely, conceptual nature, relative nature, and absolute nature.

First, science is revealing certain cause-and-effect relationships in portions of the relative world that have been separated out for the purposes of exploration. The law of gravitation, relativity, quantum physics, Maxwell's laws of electromagnetism, all of these provide cause-and-effect connections which are valid in a limited sphere. We can recognize that the field of application of *every* such cause-and-effect relationship is limited and applies to situations which have been artificially separated from the totality of all that is, and that we are therefore inevitably neglecting other causal relationships which we may not yet know but which may also be important. At the theoretical level, for example, if we neglect the finiteness of the speed of light we do not understand relativistic effects. At the practical level, if we neglect the effect of industrial waste we destroy the environment. At the personal level, if we neglect our felt relationship with our environment we destroy all meaning. However, so long as we understand and remember the importance of the larger context, then in revealing causal relationships science is dealing with relative nature. The value of science here is that it can reveal our conditioned nature *as* conditioned. It can show us the constraints within which we live.

Next, by providing positive descriptions of the world and asking ourselves and our children to believe them, science provides the belief system by which our society organizes itself. That is, it provides the ongoing conversational context by which we are bound together as a human society. It is inevitable that every human grouping have such a conversational context; in fact, it is part of what *makes* it human. Not every such social conversation has to be the same. The conversational contexts of traditional China, of medieval Europe, and of the Kung San were different from ours, but nevertheless produced genuine human societies. So long as this social conversation includes the injunction to believe in it as ultimate truth, then it becomes the conceptual nature. It becomes imaginary and a means to social and individual self-deception and thus to arrogance and aggression. If, however, the conversation remains open-ended and points beyond itself, then it remains part of relative nature and a means to discover the unconditioned nature within this.

Finally, science may bring us directly up against ineffable, unconditioned nature at the point at which error is revealed in an entire belief system. At that point, when there is a mismatch between our entire world view and *what is* then for a moment, we might be able to drop our imaginations altogether and directly experience the unlimited possibilities of what is. This was the case with some scientists at the beginning of the century as they saw the breakdown of classical physics and of an entire world view that went with it. Alfred North Whitehead, whose work we will study in Chapter 16, is a notable example. From this point of view, realization of the limitations of Newtonian mechanics or Darwinian evolution, and the extent to which we have allowed these theories to form the basis of belief systems far beyond their realm of applicability, is more important than the nature of the theories that might replace them. Realization of the falsity of the camera theory of perception (that our eye is like a camera and that seeing is like looking at a photograph inside our head) is far more important than whatever new theory someone comes up with to replace it. Realization that language ability is to a large extent a result of evolution rather than given specially to humans from some super-natural source is more important than the specific theories as to how language evolved. The replacement theories *are* important, but only insofar as later on they themselves will reveal a presupposition to be false, or will reveal the limitations of an organic disposition, until then hidden from view.

To stay with such a point of view is like trying to cross Niagara Falls on a high wire. It is so easy and so much more comfortable to begin to believe the new theories and weave beautiful new fantasies about the nature of reality—the eternalist view. Or, the other side of the high wire, it is so easy to believe that because our descriptions cannot themselves be the truth about reality, therefore there is no reality at all beyond our conceptual systems—the view of the nihilist.

In this book we will try to stay on the high wire, to take the middle way, so to speak, although we may well slip from time to time. To stay with the middle way is always to remember the metaphorical, poetic nature of language and imagery, pointing beyond what is said to nonconceptual, unbounded openness. In Part One we have estab-lished the gound, the viewpoint from which our journey of uncovering might begin. In Part Two we will undertake that journey, albeit in a small way. We will examine the belief systems about our world which arise within the sciences of biology and physics. We will find, just as we did when, in Chapter 4, we examined the history of ideas in physics, that some belief systems are being born just as some are decaying. The birth and decay of belief systems naturally points beyond such systems altogether, to the unconditioned nature which they attempt to grasp. These are belief systems which organize our society as well as our individual organisms. They are belief systems in which most readers grew up and which, because we are unaware of them, are a source of narrowing and tightness. By uncovering and working through these beliefs, the reader might occasionally catch a glimpse of a larger world.

Part Two
The Path: Uncovering Beliefs and Conditions

6   Darwin and the "Struggle for Life"

The major theme of this book is the possibility of direct perception of the phenomenal world, including the perceiver, a perception which recognizes nondual, unconditioned nature: in other words, a view of the world as fundamentally sacred. Our method is to investigate the nature of our inherent conditioning and our belief structures, which both obscure and illuminate our unconditioned perception. All perception takes place in the context of the body. It arises with the body and through the body: I see with my eyes, or perhaps we could simply say my eyes see. In either case, whether or not there is an "I," there are certainly eyes, ears, and so on. Our experience takes place in the human body, and therefore it is important, to begin with, to understand the constraints imposed on our experience by the very fact that it does take place in the context of our body.

We will begin by looking at the question of how such a body came about at all. We shall look at the theory of evolution. There is a secondary reason for examining the theory of evolution at this point in our journey, which is that this theory exemplifies very clearly the factors involved in a scientific revolution as we described them in Part One. In the 1980s the theory of evolution is undergoing just such an upheaval.

Every schoolchild knows that Charles Darwin taught the theory of evolution by natural selection. Now, this piece of cultural folklore has two parts to it: (1) the theory of evolution, and (2) natural selection, the supposed mechanism of that evolution. And it is these two ideas that we will be discussing in this chapter. I should, perhaps, warn the reader that we are entering into a subject that has been characterized by a great deal of controversy in the past decade, controversy that is marked by rancor, accusations of political or ideological prejudice, and a general sense of emotionality and irrationality. Something is clearly at stake. We have met a similar pattern several times before, epitomized in Galileo and his colleagues who refused to look through his telescope, refused to look and see. "Darwinism" is today the battleground, even though, ironically, only a hundred years ago Mr. Darwin himself was hailed, quite appropriately, as the Galileo of biology.

The fact of evolution is not in dispute except by the biblical fundamentalists who call themselves Creationists. If you invoke a creator who can do anything, then you can explain anything by saying, "That's how the creator wanted it." so that does not go very far if we wish to try to understand the world as causally interconnected and undivided. What is not in dispute, then, by the vast majority of biologists, is that the human form, and all other presently existing plant and animal species, have not always existed just as they are but have appeared over the course of many millions of years. During this time, the forms of life on earth have become more and more complex, starting with a time when there may have been only one form of single-celled organism, to the tremendous variety of complex organisms we find today. That is not disputed. That is the theory of evolution.

However, the *way* in which this evolution happened, the causal influences or, if you like, mechanisms leading to the present complexity, is a quite open question. Now, why should such a question be of interest to lay people who may or may not look to science for some guidance about their lives? The reason is that our social structure and aspirations are grounded to a large extent on a tremendous exaggeration; natural selection, a principle which gives a partial explanation of some but by no means all phenomena observed in the biological world, has been puffed up and promoted for several generations as the Great Law of Nature by which everything, including all human behavior, can be ultimately explained. Thus, again, it is not at all questioned that there are some biological phenomena that can be quite nicely explained by invoking the principle of natural selection of the fittest; and that there are also some phenomena that can be explained rather circuitously in this way. What is being very strongly questioned is that *all* evolutionary phenomena can be explained in this way or, to put it another way, that this principle is the only or even the major principle needed to explain evolution.

Now let us go back, briefly, to the beginning of the nineteenth century to see how these conceptualizations came into being. In Chapter 4 we discussed the Aristotelian legacy that Aquinas reformulated in accordance with Christian doctrine and passed on from the Middle Ages. Part of this legacy was a view regarding living forms. It was held that all possible plant, animal, human, and superhuman forms had been created at the beginning of the world by God. All these forms were arranged in a pyramidal hierarchy with God at the top. Just as earth, air, fire, and water had their natural place, so too every one of God's creations, the creatures, had its natural place and its proper function to perform there. This was the Great Chain of Being, and it was completely full, every possible creature having been created out of God's generosity. it was completely static, unchanging, and it was continuous; there were no gaps between species because every kind of being that could conceivably be created had been.

Although the Galilean revolution had swept away the Aristotle-Aquinas explanation of the physical world, this conception of the Great Chain of Being was still the dominant belief about how living beings came into existence at the beginning of the nineteenth century. It was accepted, for example, by the great systematizer Linnaeus who designed the system of classifying species we use today when we name a flower or insect with two Latin names, genus and species. Linnaeus described and named thousands of new species of plants and animals. However, because there did appear to be gaps between species, and because God had created an *unbroken* chain of being, there was already at this time a search for "missing links" between species. In particular, there was speculation about the gap between apes and men, although there was, of course, no suggestion that man had actually developed *from* apes, or that the two had common ancestry. It was a *static* world in which all species had been created from the beginning just as they are now. And the beginning was widely believed to be the year 4004 B.C.

Yet in the eighteenth century there was also speculation about the possible changes of species over time: evolution. And at the beginning of the nineteenth century several writers were proposing a theory of evolution. These included Lamarck, who later received scathing criticism for his proposal that environmental interaction could produce changes in organisms that could be passed on to their offspring (the inheritance of acquired characteristics). An important step was provided by Sir Charles Lyell who through his writing on geology convinced most people that the earth was not six thousand but millions of years old. Thus there came into current thinking adequate time for evolution to occur.

The Mechanism of Evolution: Charles Darwin

In consequence, when Darwin prepared to write his great work *The Origin of Species*, almost all the strands were available to him to formulate the theory of evolution; the idea that species (including man) have changed over the centuries, the more complex ones having gradually evolved from the simple.

Darwin was a meticulous observer, and his writings are filled with a richness of detail and example. *The Origin of Species* is regarded by some as being a literary as well as a scientific masterpiece, evoking as it does a sense of the beauty, variety, and magnificence of the natural world. The final paragraph of the book, a summary of the entire theory, evokes a feeling for the many, many hours Darwin must have spent appreciating and delighting in this world: "It is interesting to contemplate a tangled bank, clothed with many plants of many kinds, with birds singing on the bushes, with various insects flitting about, and with worms crawling through the damp earth, and to reflect that these elaborately constructed forms, so different from each other and dependent on each other in so complex a manner, have all been produced by laws acting around us. These laws, taken in the largest sense, being Growth and Reproduction; Inheritance which is almost implied by reproduction; Variability from the indirect and direct action of the condition of life, and from use and disuse; a Ratio of Increase so high as to lead to a Struggle for Life and as a consequence to Natural Selection, entailing Divergence of Character and the extinction of less improved forms. Thus from the war of nature, from famine and death, the most exalted object which we are capable of conceiving, namely the production of the higher animals, directly follows. There is grandeur in this view of life, with its several powers, having been originally breathed by the Creator into a few forms or one; and that whilst this planet has gone cycling on according to the fixed law of gravity, from so simple a beginning endless forms most beautiful and most wonderful have been and are being evolved."[1]

Darwin pointed in three main directions for evidence of the fact of evolution or "modification by descent," as it is more accurately called. First, there is the evidence of the fossil record. During the latter part of the eighteenth and early nineteenth centuries geologists had unearthed many fossils and recognized that most of them had been formed from species no longer living. At the same time, few living species were represented in the fossil record. As more and more fossils were discovered, it became evident that gradual changes in physical characteristics could be traced through time; fossils in successive rock layers, representing successive geological periods in time, were similar to each other, yet showed gradual differences over time.

Second, Darwin pointed to the occurrence in nature of examples of modification by descent. Although it is rarely possible to see such changes actually happening owing to the long time periods involved, clear indication of this principle comes from comparing the anatomy of living organisms. For example, the forelimbs of vertebrates from human, to cat, sheep, horse, bird, and seal all have remarkably similar arrangement and number of bones. The simplest and most straightforward explanation of this is that all these creatures have a common ancestry. Many species possess in reduced and nonfunctional form structures that in other species have important functions; for example, a pig walks on only two toes while two others dangle uselessly above the ground. This convinced Darwin even further of the validity of this theory. For if pigs had been created quite independently, why would they have been created with two useless toes? Third, Darwin saw the changes that breeders could artificially produce in domesticated plants and animals as strong evidence for his theory. The modern domesticated varieties of corn and tomatoes, of chickens and dogs are so vastly different from anything found living wild that one could hardly doubt that species change could and did happen.

It was this last set of observations, the changes induced by artificial selection in breeding, that gave Darwin a clue to the mechanism of modification by descent. He realized that some form of selection must be occurring in nature, but what was this mechanism? As Herbert Butterfield succinctly remarks, "The work of Malthus and the economic writings of the industrial revolution were soon to supply what was needed here."[2] It was his reading of Malthus's *Essay on the Principle of Population* that was to give Darwin his idea for the mechanism of evolution: the struggle for existence. Malthus's principle of economics was that population would continually outpace food supply unless it was kept down by elimination of the inept and poor. Darwin expanded this to become the great explanatory principle of all of nature: in each generation more individuals of a species are produced than can survive; therefore there is a struggle for existence. Those individuals whose small chance variations best fit them to the environment will be naturally selected to survive and leave offspring. Thus, these small chance variations, if they are heritable, will be passed on to the next generation, and gradually a change in the entire population will result. The whole species will gradually evolve to be a better and better "fit" to the environment. The full title of Darwin's book laying out this theory was *On the Origin of Species by Natural*

Selection, or the Preservation of Favored Races in the Struggle for Life. This is
how he summarized his theory: "As many more individuals of each
species are born than can possibly survive; and as consequently there
is a frequently recurring struggle for existence, it follows that any
being, if it varies however slightly in a manner profitable to itself,
under the complex and sometimes varying conditions of life, will have
a better chance of surviving and thus be naturally selected. From the
strong principles of inheritance, any selected variety will tend to
propagate its new and modified form."[3]

One of Darwin's most well-known exemplifications of this
process is the case of the finches of the Galapagos Islands. Darwin
noticed, first, that these finches were very similar to finches of
mainland America. Yet there were differences. There were fourteen
different species of finches on Galapagos, none of which was found on
the mainland. Galapagos was so far from the mainland that Darwin
assumed all fourteen species could not have arrived on separate
occasions. He therefore assumed that all fourteen species had evolved
from a single colonizing species. He noticed that each species had
different dietary habits: some lived in coastal areas, some in forest
areas, some in trees, one in bushes; some species were vegetarians,
others ate insects, and so on. Further, Darwin noticed that the form of
beak of each species was peculiarly well suited to its type of diet. He
concluded that the fourteen different species had arisen as small
structural variations, particularly in beak formation, that bestowed
some advantage on the possessor: possibly that of being able to make
use of a new form of food supply as competitive pressure for the
common supply increased.

This basic mode of explanation for the variations of structure and
even behavior found in living species has, since Darwin, been used
again and again. And indeed in many cases it is so highly plausible as
to be almost certainly correct. The way some animals and insects
camouflage their appearance from predators by adopting the
coloration patterns of their usual environment; the tremendous
variety of mating calls in bird species; the mimicry of poisonous
butterflies by nonpoisonous ones; the way whole populations change
with changing environmental conditions; the complex networks of
interactions between plants, animals, birds, and insects—in particular
ecological habitats and the way a small change, such as the extinction
of one insect species, affects the balance of the entire network; the
imprinting behavior of young geese, in which they become attached to
their natural mother or a surrogate mother within the first few hours of

life; all the phenomena and so many more can be understood through the principles of evolution and natural selection.

Neo-Darwinism: The Orthodox View

Darwin himself did not know the specific biological mechanism by which stable characteristics could be passed on by inheritance and by which small variations in these characteristics might arise and themselves be inherited. He appears to have thought the variations were passed on by some manner of blending of the characteristics of the parents, and that variations might arise directly owing to environmental pressures. It was at the beginning of this century that the actual mechanism was uncovered, with the discovery in 1900 of work that an Austrian monk, Gregor Mendel, had published more than thirty years previously. Mendel cross-fertilized a smooth variety of peas with a rough-skinned variety. He found that the offspring of this mixing were all smooth rather than a blend of the two characteristics. When he then cross-fertilized this first generation, he found that, of the second generation, approximately three-quarters were smooth while a quarter were fully wrinkled. That is, the wrinkled characteristics had been passed on through the first generation in a hidden or recessive form, to reappear intact in the second generation. From experiments such as these, Mendel proposed that traits such as rough/smooth were carried on units or blocks, later known as genes, which were passed from parent to offspring intact.

Experiments similar to those of Mendel were performed by T. H. Morgan on the drosophila fruit fly, and a very large number of traits were found to be heritable in a similar fashion to the smooth/rough trait of Mendel's peas. Wing size, number of bristles, ability to fly toward or away from light, egg production, and many other characteristics of form and behavior could be shown to follow Mendel's laws of genetic inheritance.

In the 1930s and 1940s, the importance of studying the genetic characteristics of whole populations rather than individuals was recognized. The characteristics of an individual can be regarded as simply the particular expression of a gene pool shared by the entire population. In this view, the distribution of various characteristics in the population can be calculated from simple assumptions about the frequency of genes in the gene pool; the way traits coded on many genes can combine to produce adaptive characteristics; and the changes in gene frequency with environmental changes. Population genetics became a powerful tool for predicting the characteristics of a population, the way they might evolve in response to changes, and the effect that particular gene changes (mutations) might have on the population. Population genetics and Mendel's principles of inheritance were brought together with Darwin's original ideas of natural selection in a new synthesis known as "neo-Darwinism," which became the orthodox view.

The orthodox view was further strengthened by the discovery that a particular very large molecule, DNA, appeared to be the carrier of genetic traits. In 1944 Avery, McCleod, and McCartney bred a pure strain of bacterium pnennoccus which had smooth cell walls. They then placed in the flask containing this pure strain some DNA extracted from a pure strain of pnennoccus having rough cell walls. Very soon, rough-celled bacteria began to appear in the flask. The only way the roughwalled trait could have been transferred to the colony of smooth-walled cells was via the added DNA. Furthermore, this DNA must have been able to enter the cells and transform the characteristics of some of them from smooth-walled to rough-walled.

In the past two decades, this simple process of DNA transformation has been refined to make possible genetic engineering in which cells can be made to take on entirely new characteristics by the introduction of DNA from different species. For example, bacteria can be made to produce insulin, usually made only in higher organisms, by introducing into them a piece of DNA carrying the insulin gene. Thus genetic engineering has commercial possibilities. It is a dramatic confirmation of the "central dogma" of molecular biology: that DNA carries the information to direct the manufacture of all the components needed to build a single cell or a multicelled organism.

In the 1950s and 1960s, the actual structure of the DNA molecule and the way in which it directs the manufacture of proteins, the vital components of cell metabolism and structure, were unraveled. DNA is a very long string of thousands of subunits. There are only four different kinds of such subunits, usually labeled A, C, G, and T, and the order of these subunits along the DNA molecule constitutes a code—the genetic code—which carries information for directing the synthesis of the components of the cell. In directing the synthesis of a specific protein, the corresponding section of DNA is copied onto another molecule—the "messenger." This "messenger" then travels to another large group of molecules (a ribosome) whose function is to translate the code from the messenger and build the corresponding protein. Thus there is a linear, one-way flow of information from the chromosome, as the long DNA molecule is called: gene to messenger to protein to inherited characteristic. In practice, of course, the situation is not nearly as simple as this model would imply. Even such a clearly distinguished characteristic as eye color involves the cooperation of many proteins, and therefore many genes contribute to the characteristic. In a particular cell at any time, not all genes are directing protein synthesis; they can be turned off and on by "regulator" molecules. In higher organisms, only a small fraction (20 percent) of the DNA appears to be involved in protein synthesis at all; the remaining 80 percent of the genes are concerned with regulatory functions. The individual genes, far from being the static carriers of information, are seen to be in a continual dynamic state, duplicating, moving around the chromosome, and sometimes even splitting up so that a piece of DNA coding for one particular protein may be found spread out in several places around the chromosome. Nevertheless, the simple one gene/one protein model is the basic, functional building block out of which the metabolism and structure of the entire cell or higher organism could in theory be constructed. DNA, then, provides a molecular basis for understanding how small variations can arise and perpetuate themselves through alterations in the genetic code. Thus the orthodox, neo-Darwinian view appears to be broadly substantiated at the molecular level.

Now let us review what are the main assumptions in this view. First, the *only* factor taken into account for the poorer chance of survival of one member of a species relative to others is its poorer fit to the environment. The environment here includes other species and other members of the same species. Thus, the environment is viewed as essentially hostile, and the individual's relationship to its species brothers and sisters, essentially competitive: Darwin referred to "the war of nature and the struggle for existence." Second, variations *all* arise by chance. Thus the environment, apart from *requiring* that organisms "fit" it, does not participate in promoting the required variations (this would be a version of Lamarckism). Nor can there be mechanisms within the organism that promote variations in specific directions (internal needs being worked out over larger time periods). Third, *all* such changes occur in imperceptible gradations. There are no sudden jumps or "catastrophes." Fourth, *all* changes that become fixed, that is, passed on to future generations, do so in this way. This, then is the orthodox view which is now being challenged.

Is Neo-Darwinism The Final Truth?

Perhaps we should pause at this point to appreciate the tremendously liberating effect that Darwin's ideas had on Western man's world concept. An eternally static world in which organisms simply reproduce embodiments of the ideal form already created at the beginning, apparently arbitrarily by a Creator, became ever-changing, continually bringing about a seemingly inexhaustible variety of new forms by processes inherent in nature. Man himself, from being a creature of fixed nature divinely created for the incomprehensible purposes of the Creator, always separated from and antagonistic to the natural world, became a part of that world, evolved and continuing to evolve according to the same inherent laws by which all of nature evolves, connected in the same vast and complex web of interactions as all of nature. Just as, with Copernicus, humanity was opened to the vastness of the universe, so with Darwin humanity was opened to the inexhaustible creativity of nature. Just as, with Galileo, earthly and heavenly laws of motion were unified, so with Darwin the physical constitution of men and women was now to be understood according to the same laws as all of nature.

However, predictably, just as Newton's laws were bloated into metaphysical principles as to the final nature of time, space, and matter, so natural selection has been bloated into a principle of explanation of phenomena far beyond its scope. The idea that nature was fundamentally based on a "struggle for life" and that man was part of that struggle was promoted into a social philosophy and an explanation of human nature that was at the very heart of the industrial revolution and that continues today in the world of business, economics, education, and international relations.

To quote Herbert Spencer, a popular nineteenth-century philosopher, "the wellbeing of existing humanity and the unfolding of it into this ultimate perfection, are both secured by the same beneficient, though severe discipline, to which the animate creation at large is a subject. The poverty of the incapable, the distresses that come upon the impudent, the starvation of the idle, and those shoulderings aside of the weak by the strong, which leave so many in 'shallows and miseries' are the decrees of a large far-seeing benevolence."[4] Again, Thomas Huxley, perhaps the man most responsible for bringing evolution to public knowledge: "Among primitive men the weakest and stupidest went to the wall, while the toughest and shrewdest, those best able to cope with their circumstances, but not the best in any other sense, survived. Life was a continual free fight, and beyond the limited and temporary relations of the family, the Hobbesian war of each against all was the normal state of existence. So long as the natural man increases and multiplies without restraint, so long will peace and industry not only permit, but necessitate, a struggle for existence as sharp as any that went on under the regime of war."[5] More recently, John D. Rockefeller, Sr., said succinctly, "The growth of a large business is merely a survival of the fittest. . . . It is merely the working out of a law of nature and a law of God."[6] And this view is continued today, as we see in the most recent piece of extravaganza by Richard Dawkins, author of *The Selfish Gene*, "We are survival machines—robot vehicles blindly programmed to preserve the selfish molecules known to us as genes."[7] These men themselves were not speaking out of ill feeling particularly, but they were spokesmen for the philosophy of the age, a philosophy which is still deeply rooted in our culture.

The nature of the jungle may have changed in some nations from a material to a psychological or spiritual one, but the same principle still guides our society. In a nutshell, it is the principle that the world, particularly nature, is essentially hostile, that human nature is essentially selfish and cannot be changed, and that genuine, egoless compassion for others is impossible. "Natural selection" is claimed to prove this, and it is taught in every high school and undergraduate biology or general science course throughout the modern world. Yet this doctrine is controversial and suspect to many of today's leading biologists. Here is what one of the world's greatest living biologists, editor of a twenty-eight-volume treatise on zoology and ex-president of the French Academy of Sciences, Pierre Grassé, has to say about natural selection: "Their success among certain biologists, philosophers and sociologists notwithstanding, the explanatory doctrines (natural selection) of biological evolution do not stand up to an objective in-depth criticism. They prove to be either in *conflict with reality or else incapable of solving the main problems involved.*

"Through use and abuse of hidden postulates, of bold, often illfounded extrapolations, a pseudoscience has been created. It is taking root in the very heart of biology and is leading astray many biochemists and biologists, who sincerely believe that the accuracy of fundamental concepts has been demonstrated, which is not the case."[8]

Given the tremendous confirmation of evolution that we have sketched out, what would bring such a prominent biologist to make such comments about the doctrine at the foundation of his own discipline? In discussing the theory of evolution by natural selection, it is important to understand that we are using the term "theory" in a rather different way from its use as in "theory of gravitation." The theory of natural selection is not confirmable or refutable in any way, as Karl Popper has shown. It is a principle that produces plausible explanations, but it has not and cannot be "proved." "Neither Darwin nor Darwinianism has so far given an actual causal explanation of the adaptive evolution of any single organism or any single organ. All that has been shown—and this is very much—is that such an explanation might exist—that is to say they are not logically impossible."[9] To quote British zoologist Leonard Matthews, "The fact of evolution is the backbone of biology, and biology is thus in the peculiar position of being a science founded on an unproved theory—is it then a science or a faith?"[10]

Consider again the fossil record. If Darwin's notions of gradual evolution in small, almost imperceptible steps were the full story, then we should expect to find such small steps joining various fossil forms in finely graded series. We do not. As Darwin himself said, "The number of intermediate varieties, which have formerly existed on earth, must be truly enormous. Why then is not every geological formation and every stratum full of such intermediate links? Geology assuredly does not reveal any such finely graduated organic chain: and this perhaps is the most obvious and gravest objection which can be argued against my theory."[11] Darwin hoped that as the years passed these gaps in the fossil record would be filled in, but they have not. Professor Herbert Nilsson of Lund University, Sweden, sums it up this way: "It is not even possible to make a caricature of evolution out of palaeological facts. The fossil material is now so complete that the lack of transitional series cannot be explained by the scarcity of the material. The deficiencies are real, they will never be filled."[12]

There are a number of phenomena which are either simply ignored by the natural selectionists or explained in arguments so convoluted that they could be regarded as plausible only with an act of faith. First, as we have seen, evolution has to be exemplified in the fossil records. These records certainly do confirm the bare fact of evolution. But far from confirming the idea of small gradual changes linking all species to a common ancestor, the fossil records now unequivocally *do not* confirm this. For example, it is said that amphibians appeared when fish began to move onto the land. There should therefore be transitional fossil forms in which fishes' fins are on the way to becoming limbs. There appear to be no such transitional forms that are at all plausible.

The same is the case with the origin of flight: there simply are no intermediate fossil forms between birds and land-borne reptiles. And how could a creature half fly? Its system of flight would have to be functional or a useless encumbrance. The origin of insects and flowering plants is also almost a complete mystery. These are not simply unresolved issues that will clear up over the course of time, making natural selection theory an exciting area for research. They are fundamental obstacles to the theory of natural selection, similar to the negative result of the Michelson-Morley experiment which gave rise to the theory of relativity.

Next, there is the question of the evolution of complex systems: the eye, the ear, the mammalian jaw, blood circulation, poison apparatus in snakes and insects. The list could be much extended. They are all examples of complex structures involving many mutations which must all work together in almost perfect harmony or the structure will be utterly useless. For example, in the eye, to take only the major aspects, there is the bone structure of the eye socket, the lens, tear glands, and ducts, muscles of coordination, the rods and cones of the retina and their connection to the optic nerve, the specialized areas in the brain for processing optical information: all of these parts must be functioning together for there to be vision. If any one of these had appeared, randomly, without the others, it would have been useless and, according to natural selection, selected out. As Stephen Jay Gould says, "What good is half a jaw or half a wing?" Yet the probability that *all* these changes occurred *randomly* at the same time, to produce a mutant with a seeing eye, is too low to be thinkable. Similar arguments apply to the other complex systems we mentioned, and many more.

Broadening The Orthodox View

Recently some plausible possibilities have been proposed as ways in which our understanding of evolution could be broadened somewhat beyond the orthodox view.[13] First, in order to understand the suddenness of changes, some older ideas known as "catastrophes" and "hopeful monsters" are coming back into use. The idea of "catastrophes" or "punctuated equilibrium" is that for long periods of time, while the environment was relatively stable, there was very little evolution of species. But at rare intervals, huge natural catastrophes, such as large-scale geological shifts or the striking of the earth by a sizable meteor, caused very widespread destruction of existing species. Such catastrophes could also be internally produced. That is, the entire biosphere might at times have reached a critical point at which total collapse occurred. We will see that this is a possible response of a complex system to a large energy input in Ilya Prigogine's work on open systems. The mathematical theory of such catastrophes has also been worked out by René Thom. In response to the drastically changed environment, radically new species appeared. "Hopeful monsters" refers to the way in which such new species may have come rapidly into being. Perhaps in response to the tremendous environmental pressures after the catastrophe, a large number of grossly mutated organisms appeared (monsters), most of which died, but some of which were better adapted or rapidly able to adapt to their new environment. Such mutations would be most likely at the embryonic level, affecting a number of coordinated changes in the embryonic development.

Scenarios of this type are equally plausible but somewhat contrary to orthodox neo-Darwinian theory. This is especially so since a *relaxation* in competitiveness would be more likely to allow the "hopeful monsters" to experiment with various forms of behavior, diet, etc., and possibly therefore survive. Furthermore, *cooperation* between potential survivors would seem to be most helpful. We should remember that orthodox neo-Darwinism depends on an environment in which resources are limited relative to a particular population—a situation which would not necessarily be the case after a catastrophe.

A further possibility that would help us to understand these sudden rapid changes is a revival of the interaction between genetic inheritance and behavior. If the organism adopts certain new behaviors, such as seeking a new habitat or trying a new diet which enables it to survive, then *following* this change of behavior the pressure of the environment would result in the selection of offspring who are better adapted on the genetic level. This is known as the Baldwin effect and has been observed in nature. Konner gives the following example: "Consider a population of birds in which some individual learns to like a new form of berry, say blueberries. These individuals start nesting in blueberry patches and their offspring learn to like blueberries just as they did. Eventually, just randomly, the genetic shuffle produces a few individuals who like blueberries right off—they don't have to go through the process of learning. These individuals may be favored by selection and may reproduce so effectively that eventually we observe a generation in which all individuals have the genetic propensity to like them without learning."[14] This illustration may be extended to less innocuous cases; for example, instead of birds learning to like blueberries, we might be speaking of humans learning to like cooked food, or war. It is highly plausible that learning has, in the manner of the Baldwin effect, been a major factor in evolution from the most "primitive" organisms up to man. Thus the language of behavior could be quite appropriate for describing evolutionary change.

This theory is in practice indistinguishable from Lamarckism, except that the mechanism is not via *direct* influence of behavior on genes in one generation. However, it appears now that there may be a mechanism for direct Lamarckism in certain situations. It is known that sex steroids combine with reception molecules in target cells and that the steroid-receptor combination then has a direct influence on DNA. This means that it is quite possible that persistent behavior that affects the levels of, for example, testosterone, such as diet, stress, sexual activity, could conceivably affect the DNA of the germ cells. For example, it could be that exposure to certain levels of testosterone could permanently unlock certain genes in the DNA in all the body cells including the germ cells. Thus, the behavioral tendencies brought about by unlocking these genes could be heritable. We do not know that this happens, but the point is that we now have a known, perfectly acceptable mechanism for the most strict form of Lamarckism, or inheritance of acquired characteristics. We should remember that in higher animals only about 20 percent of the DNA is involved in directing protein synthesis and therefore determining traits directly. The remaining 80 percent is presumed to be involved in control functions directing which traits will be expressed and which repressed at particular times in the life cycle of the organism. It is therefore highly plausible that sections of the DNA which may have been repressed for generations, and therefore inherited as nonfunctional genes, may become "turned on" by environmental conditions and subsequently inherited as functional genes.

The theories of "catastrophe," "hopeful monsters," and gene-behavior interaction are really rather natural extensions of the neo-Darwinian theory, and from the point of view of biology itself it is quite surprising that so much furor has arisen around them, a furor which has strong political and religious overtones. Lamarckism was for decades endorsed by Marxist theorists, since it seemed to confirm the idea that a drastic change in the environment due to revolution could, within one generation, bring about drastic changes in inheritable characteristics. The catastrophe theory leaves an open door to the idea of divine intervention, for the catastrophes may have been divinely rather than naturally caused. These factors are often involved in any debates on evolution, and it is quite probably such hidden agendas that prompt the emotionality of the debates. The theories themselves are rather innocuous. They enlarge the orthodox view enough to explain the gaps in the fossil records, the sudden appearance of new species and extinction of others, and the fact that some primitive forms escaped selection. But they are still largely adaptionist; that is, they still take the view that natural selection of the best-adapted organisms is the final arbiter of what survives and what does not, after a catastrophe or as a result of environment-behavior interaction. And the confusion about this doctrine amongst even the proponents of the orthodox view is clear. Here are two statements, both written in 1980, each by highly regarded neo-Darwinists:

1. "The theory of evolution is based on the struggle for life and the survival of the fittest."—Robert Axelrod and Wilbur Hamilton[15]
2. "In presenting his theory, Darwin made a serious mistake in characterizing natural selection as a 'struggle for life.' To what extent does natural selection depend on the outcome of violent struggles or lethal combat? The answer is, very little."—Ledyard Stebbins[16]

This, then, is the dilemma of modern biology. The theory of evolution, through the genius of Charles Darwin, swept away the narrow vision, bigotry, and dogmatism of nineteenth-century biology and religion. Yet natural selection, the keystone of that theory, has now itself become the source of narrow vision and dogmatism, both in biology and in the whole fabric of our society.

The view that is propagated in schools, universities, and popular magazines throughout the modern world, namely, that the one and only explanatory principle of nature, including human nature, is "natural selection and the struggle for survival," is not an accurate description of nature. And yet a great many biological, sociological, anthropological, psychological, economic, and political theories are based on this view and have been used to direct all kinds of practical affairs.

We are in a situation with regard to evolutionary theory very similar to that in physics at the end of the nineteenth century. The pressure on the orthodox theory by clear, incontrovertible observations not fitting that theory is too great. There *must* be a broader theory in which natural selection could be incorporated as one, perhaps major, part.

I should add that this certainly does not mean that human behavior is not based on and constrained by the biological functions of the human body, and we will see later what some of these constraints might be. It does mean, however, that we have to be very circumspect and cautious in examining such claims, to see how much is based on actual observation and how much is based on the "pseudoscience," to use Grasse's term, of natural selection. In particular, the view that egocentrism is a biological necessity is based on this pseudoscience. Egoism may be strongly reinforced by deep-rooted genetic factors in our biological being, as we shall see. But it is not a biological *necessity*. In the next chapter we shall see that the aspect of mutual cooperation between organism and its environment and "kindness" toward others may be a principle of explanation as fundamental as natural selection by survival.

7 Mutual Cooperation

In the history of so many of the ideas of science, that cyclical process in which, for a while, a partial truth is proclaimed as the one "big truth," is always accompanied by dissenting voiced presenting an opposing or a balanced view. This is no less true of natural selection. One such dissenting voice was that of Petr Kropotkin who in 1902 published a book called *Mutual Aid*. Shortly after the publication of *The Origin of Species*, Kropotkin spent five years in the field in Siberia. He was honored for his work there with a gold medal from the Russian Geographical Society. As he says, he eagerly looked for "that bitter struggle for the means of existence, among animals belonging to the same species, which was considered by most Darwinists as the dominant characteristic of struggle for life, and the main factor of evolution."[1] However, he came to the conclusion that an equally significant factor was mutual cooperation and assistance among animals.

Kropotkin's book is replete with examples of this. Here is one particularly delightful one, in which the straightforward explanation is that the animals were helping another in distress: "As to the big Molucea crab I was struck [in 1882 at the Brighton Aquarium] with the extent of mutual assistance which these clumsy animals are capable of bestowing upon a comrade in case of need. One of them had fallen upon its back in a corner of the tank and its heavy saucepan-like carapace prevented it from returning to its natural position, the more so as there was in the corner an iron bar that rendered the task still more difficult. Its comrades came to the rescue and for one hour's time I watched how they endeavored to help their fellow-prisoner. They came two at once, pushed their friend from beneath and after strenuous efforts succeeded in lifting it upright, but then the iron bar would prevent them from achieving the work of rescue, and the crab would again fall heavily on its back. After many attempts one of the helpers would go in the depth of the tank and bring two other crabs, which would begin with fresher forces the same pushing and lifting of their helpless comrade."

Again, in a book written in 1931, we read, "What has got into circulation is a caricature of nature, an exaggeration of part of the truth. For while there is in wild nature much stern sifting, great infantile and juvenile mortality, much redness of tooth and claw . . . there is much more. In face of limitations and difficulties, one organism intensifies competition, but another increases parental care; one sharpens its weapons, but another makes some experiment in mutual aid. . . . The fact is that the struggle for existence need not be competitive at all; it is illustrated not only by ruthless self-assertiveness, but also by all the endeavors of parents for offspring, of mate for mate, of kin for kin."[2]

Darwin himself was quite well aware of the many examples of cooperative behavior amongst animals that can be seen in nature. With the emphasis on populations in the neo-Darwinian synthesis, mutual cooperation was more or less dismissed as merely a trait which obviously furthered the survival of the group. However, in the 1960s interest turned back to the individual and to the question of how traits that, while favoring the group, may not favor individuals, could possibly survive in the gene pool. For example, in a flock of birds, the bird which, on detecting a predator, cries an alarm call warning the whole flock would, one would expect, be the most vulnerable to the predator's attack. Thus one would expect that if such alarm calling is genetically based, the genes carrying the tendency to call would be rather rapidly selected out. Nevertheless, there are many examples of such helping behavior in nature.

Kin Selection

It is now clear that at least some of this cooperative behavior is genetically based. This is the idea of "kin selection," which holds that it is genetically favorable for closely related animals to act in a mutually cooperative way, thus providing their family genes a higher chance of survival. For example, a pair of birds may be assisted in the care and feeding of their young by one or more younger birds, so that three or four adults are taking care of the nest. This has been observed, for example, in the Florida scrubjay. This genetically based tendency for kin to assist and protect each other is also known as "inclusive fitness," referring to the fact that the idea of "natural selection of the fittest" must now be extended to include close family. That is, new patterns of behavior become established if they enable certain individuals *or families* to produce a larger number of offspring. "Producing offspring" implies not merely giving birth to them but also feeding and protecting them, possibly for many months or even years. This is the genetic basis, then, of mutually cooperative behavior.

One could interpret the phenomena of "kin selection" as an example of "selfishness"—that the organism is acting "selfishly," not for its own sake but for the sake of certain genes. This is the approach of sociobiologists taking the orthodox approach such as Edward Wilson, author of *Sociobiology: A Modern Synthesis*,[3] and Richard Dawkins, author of *The Selfish Gene*.[4] However, this is a matter of interpretation, which must be based on philosophical assumptions entirely outside the realm of biology. One could just as well say that "kin selection" implies an inherent tendency in the whole organism to help those that he regards as "kin," and this is the way we will view it in this chapter. We regard this view, while still being interpretative, as being somewhat more realistic from the point of view of real experience. The organism, especially the human, usually *experiences* itself physically as a unit and acts toward others accordingly. It does not experience itself as a collection of genes fighting individually for their survival. As an individual it might experience a struggle for survival and a feeling of aggression, or it might experience a feeling of friendliness toward kin. Both have a biological basis.

Now the question arises, how can an animal, say a bird or an insect or a lion, know who his close kin are? Two young lions growing up in a pride may know each other individually, and be able to recall this after a long separation, but is there any mechanism by which they know their precise genetic relationship? Does a songbird father know which of the young birds in his nest are "his"? The answer to these questions seems to be no. There is strong evidence that there is no inborn trait enabling an animal to recognize precisely his genetic kin. Rather, animals appear to behave toward each other *as if* they were kin if the situation warrants it. This is the basis of animal societies and indeed of "cultural" behavior in animals. It also presents a profound problem for the "selfish gene" proponents.

Animal Societies

In order to understand how mutual cooperation may lead to the formation of societies, we must first distinguish between two kinds of inheritance: genetic and cultural. Genetic inheritance refers to traits that are passed on through the genes. Cultural inheritance refers to traits that are passed on in other ways, particularly through imitation, instruction, and communication. Either or both of these kinds of inheritance may be involved in the transmission of behavioral traits from parents to offspring and the formation of animal societies. Animal societies can be found at all levels of complexity, and, in particular, they are often divided into four main groups: (1) single-celled bacteria and slime mold which, in certain phases of their life cycle, can band together to form what appears as a single multicelled organism; (2) insects such as wasps, bees, ants, and termites which exist in highly organized colonies: (3) vertebrates from birds to primates; (4) humans.

As an example of the first kind of society, consider the slime mold. The individual cells are amoebas of microscopic size which live on bacteria and, so long as there is food available, reproduce by simple division. However, when food grows scarce, many amoebas join together to form a body visible to the naked eye. This now behaves like a single primitive organism. It has a front and a rear end and can slither in the direction of heat or light. After a week or two it transforms into a fruiting body with a base and a stalk. At the tip of the stalk, which is several millimeters high, spores are found which, when environmental conditions improve, can burst out to form new amoebas. At this level, the organization is a genetically fully determined response to environmental conditions. There is clearly no cultural or learned transmission involved.

At the next level, that of insect societies, there are complex patterns of social hierarchy, behavior, and division of labor. According to E. O. Wilson, there are three aspects to the most highly organized insect societies: (1) a cooperation between individuals in the colony to take care of their young; (2) a division of labor in connection with reproduction (that is, there are sterile castes—the workers and soldiers); and (3) an overlap of generations so that offspring assist their parents. With regard to the division of labor, individual insects take on a particular body size and shape and corresponding behavioral traits as they develop from the larvae. It appears that they take on these characteristics partly in response to chemical messages or nutritional conditions, and also in response to an established dominance hierarchy. There are no separate genes determining which particular insect will be a worker and another a soldier. As in the case of slime molds, however, there is no cultural transmission; all the complex behavior patterns appear to be genetically determined. This is the case even with the famous dance of the bees, in which bees returning to the hive communicate through a complex dance the distance and direction of a source of honey. The only source of flexibility in this dance is in the indication of distance and direction. The dance itself and the particular symbolism involved are genetically determined. There is some learning involved, in that the young honeybee has to learn in which direction the sun moves. However, this learning is not culturally transmitted. Thus, the two lowest levels of animal society are characterized by a high degree of specific, genetically determined organization and probably no cultural transmission. Nevertheless, it is thought that the evolutionary basis of the formation of such societies is kin selection.

When we consider the third level of society, the vertebrates up to the nonhuman primates, the picture is quite different. We find that the degree of organization of such societies is far less. At the same time, the genetic inheritance seems to have provided for much greater flexibility of behavior; that is, the range of possible behavioral responses to a particular environmental situation can be quite broad. In addition, there is the possibility of learning, so that cultural transmission of behavioral traits becomes possible. John Tyler Bonner has identified five categories of cultural transmission in nonhuman vertebrates: 1) behavioral patterns of physical dexterity; (2) relations with other species; (3) auditory communication within species; (4) complex geographical locations and routes; (5) inventions or innovations.[5]

An example of the cultural transmission of physical dexterity, adduced by Bonner, is seen in the oyster-catcher, a shorebird which feeds on the mussel. The mussel has two hard shells which are closed by a strong muscle. The oyster-catchers can use two different methods to get to the flesh of the mussel; they either put the mussel on some hard ground and hammer hard on the shell with their beaks to break it, or they wait until the mussel opens and then quickly snip the muscle which keeps the shell closed. The young oyster-catchers have to learn one or another of these methods from their parents. An adult bird was never observed to use both methods. However, if the eggs from the nest of a hammerer are transferred to the nest of a snipper (or vice versa), the young birds will learn to snip (or hammer) like their surrogate parents. The conclusion suggested is that the possibility of hammering or snipping is genetically transmitted while the learning of a particular technique is culturally transmitted. That it is indeed a learning process is further suggested by the fact that with oyster-catchers living in localities where there is a more readily available food supply, such as worms, the young stay with their parents for only six to seven weeks, while the mussel-eating young stay with their parents for eighteen to twenty-six weeks.

Another example in which there is clearly cultural transmission, although it goes far beyond simple physical dexterity, comes from Harry Harlow in his research on rhesus monkeys. Harlow found that female infant monkeys raised apart from their mothers, even though well cared for, become poor and negligent mothers when they grow up. Stebbins comments, "Even in these monkeys the genes do not by themselves ensure the development of normal maternal behavior. Their action must be supplemented by a separate thread of cultural transmission that each successive mother learns from her mother and passes on to her daughter."

Examples of cultural transmission of relations with other species have to do with learned avoidance of predators, particularly tameness or aggressiveness toward man. An example is that many birds and animals on the Galapagos Islands show absolutely no fear of man while on land. When these same animals are swimming they become quite terrified if approached by man. Because this behavior is variable and can change in one generation, we can assume that it is learned. Learned transmission of bird songs is an example of the third category, and Bonner cites the traditional use of specific routes for migrating birds, as an example of the fourth. These routes are long and complex and it is very difficult to see how knowledge of them could be genetically programmed.

A most dramatic and well-known example of learned innovations is the case of a macaque monkey on the island of Koshima, named Imo. Japanese scientists regularly put sweet potatoes onto the beach to attract a troop of monkeys they wanted to observe. Imo started to wash the potatoes to remove the sand before eating them. Soon a few animals began to imitate her, and within five years 80 percent of the macaques were washing their potatoes before eating them. Later the scientists put wheat on the beach for the monkeys to eat. Imo learned to separate the wheat from the sand by tossing a handful into the water—the wheat floated and the sand sank. Soon this too became a common activity.

Interaction between cultural and genetic transmission has been invoked especially to explain the very rapid appearance of cultural behavior in animals and man.[6] The suggestion is this: the genes prescribe certain *general* (genetic) tendencies of development, such as Chomsky's innate rules of language acquisition, or very broad learning capabilities. As the individual grows, he or she absorbs parts of the culture already in existence, according to these genetic tendencies. The culture is created anew in each generation by the mutual cooperation, decisions, and innovations of all members of the society. Some individuals possess genetic tendencies enabling them to survive and reproduce better in the contemporary culture, and these more successful tendencies are passed on through the population together with the genes encoding them, thus reinforcing those tendencies in the culture.

Notice that although such a proposal is speculative, it does go beyond traditional neo-Darwinism in utilizing gene-culture interaction. It is also a valuable model in that it indicates how genes might determine certain general tendencies of behavior while still leaving open the possibility of individual cultural influence and modification of such inherited tendencies. Wilson and Lumsden apppear to believe that *details* of cultural behavior such as religious ritual are genetically determined in this way. This seems rather fantastic, and comes from their fundamentalist view that *all* human behavior can be understood as an extension of neo-Darwinism. Nevertheless, the model of gene-culture interaction they present has some interesting possibilities. For example, does a culture which over many generations encourages aggression or, alternatively, gentleness, result in individuals who are genetically more prone to aggression or gentleness? For example, is the philosophy of egoism, which is strong in contemporary culture partly as a consequence of the widespread misapplication of the theory of "survival of the fittest," likely to result in the selection of more individuals with a genetic tendency to greater egoism? This might presumably follow from the gene-culture interaction idea. We will return later to the question of the biological basis of such basic parameters of behavior as aggression or gentleness.

The Biology of Behavior in Man

The question of whether or not, and to what extent, human behavior is determined by our biology is a controversial and lively one, with various factions endorsing various extreme views. These views range from the belief that *all* of our behavior is, or will ultimately be, explainable on a genetic basis and that no other principles of explanation will be needed, to the belief that biology, psychology, and spirituality deal with entirely different realms and that each has nothing of value to say to the other. The first view is held by those who uphold the banner of reason and science, and the second by those who sing the song of intuition. The arguments turn particularly around the relation between kin selection and a rather vague concept of "altruism."[7] However, neither of these beliefs seems to be strictly in accordance with what we can see before us, namely, that our behavior is indeed very strongly conditioned and constrained by our biological nature and that some of these constraints are genetically determined. However, while the biological constraints provide guidelines and tendencies for behavior, they do not determine what particular act will be performed by a particular individual at a particular time.

The outlines of the main known biological constraints—the nervous and endocrine systems—have been documented in careful and extensive detail by Melvin Konner in *The Tangled Wing: Biological Constraints on the Human Spirit.* This is a brilliant and heartfelt book in which Konner not only elucidates these constraints as far as we now know them, but, from time to time, with the words of a poet, shows that through the rediscovery of the sense of sacredness and awe these biological energies may be the foundation not for degradation and despair but for humaneness and dignity.[8]

The nervous system can be regarded as an integrated, self-organizing system built in a circular fashion, whose components are various types of single cells generally called neurons. A typical neuron has a main cell body with many short fibers extending from it, known as dendrites, and one long fiber, the axon, which may itself have many branches. Communication between neurons proceeds when an electrical potential is generated in the cell body of one neuron and passes down its axon in a coded sequence. When the signal reaches the end of the axon, it is ready to pass on to a dendrite of another neuron. Between the axon of one neuron and the dendrite of the next, there is a gap, the synapse, over which the message from the first neuron must pass. This is achieved via chemical transmitters known as neurotransmitters. Neurotransmitters may be of two types—excitators, which facilitate the development of an electric current in the receiving cell, or inhibitors, which inhibit the development of such a current. Any neuron may have up to ten thousand dendrites and thus be receiving excitatory or inhibitory messages from that many other neurons at any moment.

In the cortex of the brain alone, there are fifteen billion such neurons, and since any of them could be firing at any time, there are an almost inconceivably large number of neuronal interactions taking place in the brain at any moment. In the gray matter of the brain (neocortex), in a portion the size of a pinhead, there are between thirty and one hundred thousand neurons, joined by about a mile of axons. It is clear that we can only begin to approach an understanding of such a delicate and complex system by study of its subsystems and their interactions and by considering the integral states of these subsystems. It would be out of the question to try to understand the brain's functioning from consideration of the firing of individual neurons.

However, our discussion indicates the tremendously complex environment into which an incoming impulse from, say, the retina via the optic nerve enters. In particular, there are connections from the various early sensory areas to the limbic system, a structural area necessary for a range of emotional responses, and from there to the frontal lobe which is known to be involved in "foresight and concern for the consequences and meaning of events." Thus, we can already begin to see that at this biological level, there is an intricate connection between perception, action, and meaning. We will examine this in detail in Chapter 12.

Beyond these connections between the major subsystems of the nervous system, the system is constantly bathed in a continually changing and wide array—we could almost say an atmosphere—of hormones, neurotransmitters, and other brain chemicals, provided by the endocrine system. All of these affect the functioning of the various subsystems. The production of all of these substances is genetically determined, so that a genetic deficiency can lead to absence of some crucial hormone or neurotransmitter and thus to behavioral abnormalities. Nevertheless, such effects are reversible by appropriate changes in the environment. Likewise, the effects of genetic deficiencies can be mimicked in normal individuals by environmental means. Thus, we see the intricate relationship between environment and behavior at the biological level.

Individual Acts are Unpredictable

A definite pattern begins to emerge in the work of behavioral biologists. This is typically exemplified in the way Konner has titled his chapters: in the chapters titled "Joy," "Love," and "Grief," he in fact talks about the biology of pleasure, attachment, and depression. Now we know from our own experience that joy is not *necessarily* the same experience as pleasure, nor love the same as attachment, nor grief the same as depression. It begins to seem likely that our lower, preconceptual biological structures provide a range of available energies: rage-lust-fear, pain-pleasure-depression, tension-relaxation. These energies may be called forth by particular environments, and which energy is called forth and what interpretation we put on that energy depend on higher, conceptualizing functions. We cannot ignore the intimate involvement of these primary energies in every perceptual act, but, at the same time, the way these energies enter and organize the perception depends on the conceptual framework of that perception. And beyond this there is the changing, moment-by-moment interpretation of our own response, which then has a changing effect on our perception itself.

One example of this, to which all behavior could be analogous, is the relation between the utterance of a particular sentence in an actual situation, the speaker's inner state (mood, etc.) at the time, his knowledge of the language, the genetic determinants of his ability to learn a language, and the environmental situations calling forth the utterance. These five elements will be involved in any action: (1) the act itself; (2) the inner state (thoughts, emotions, bodily state, and state of awareness) of the person; (3) the cultural training of the individual, further favoring certain tendencies; (4) the inheritance of the individual, predisposing him or her to certain inner states and to respond in a certain way; (5) the environmental situation to which the action is a more or less appropriate response. These are all interconnected. Even the act itself takes place over a finite time and can therefore be influenced in midstream by changes in environment or in inner state. And, in turn, the act can affect environment or inner state in midstream. Even which particular aspects of inheritance or cultural training are predominant factors may change in the middle of the act. Thus, in order to get to the heart of the question as to whether and how we, or awareness, may influence our behavior, we have to consider the fine details of action and perception as a process in time.

This raises the question: at what point in the perceptual process does the bare, almost physically felt, biological energy of pleasure-pain, tension-relaxation, anger-lust become a definite form recognizable as an "emotion," motivating and coloring action? Another related question is: at what point in the perceptual process does one specific form of energy, one specific emotion such as anger, turn into another? For we are probably all familiar with the experience of anger or lust's becoming humor, of joy that turns into sadness, enjoyment into boredom, and so on. These seem to be "branch points" in our experience, and it is precisely at these points that a new form of energy might come in, perhaps simply awareness, to open up the choices available at such a branch point. Such questions bring us very close to understanding why the training of awareness might be helpful: usually in the hurry of everyday life we miss such branch points, and by the time we notice our internal state, that moment has passed. The reason is that our ordinary, untrained attention is not able to focus on time intervals small enough for us to be aware of the actual sequential processes of perception and therefore to be present at a branch point between, say, anger and humor, or fear and fearlessness. In a later chapter, we will look more closely at the specific process of perception.

It has been suggested quite cogently by physiologist H. B. Barlow and animal behaviorist N. K. Humphrey, among others, that self-consciousness, or at least some aspects of it, may have an evolutionary function. Barlow argues that consciousness arises primarily in the relation between one individual and another and is not a property of the brain in isolation: "Nature has constructed our brains so that, first, we seek to preserve individual consciousness; second, we can only achieve it in real discourse or rehearsed future discourse; and third, important new decision requires the sanction of consciousness. Thus the survival value of consciousness consists of the peculiar form of gregarious behavior it generates in man."[9]

Humphrey's view complements this: "For man and other animals which live in complex social groups, reality is in large measure a "social reality." No other class of environmental objects approaches in biological significance those living bodies which constitute for a social animal its companions, playmates, rivals, teachers, foes. In these circumstances the ability to model the behavior of others in a social group has paramount survival value." Humphrey argues that a function of consciousness lies in that by becoming conscious of our own sensations we may understand the sensations of others by our own and therefore predict their actions. Of course, ordinary introspection, as well as the deeper insights of sitting meditation, tells us that what is being talked about here is only one of many functions and levels of "consciousness," that dualistic process of consciousness which splits the world into "I" and "other." Nevertheless, it does seem that the possibility of sympathetic imaging of another's needs is biologically functional.

Aggression, Kinship, and Gentleness

In the work of animal behaviorists, sociobiologists, and evolutionists we find a tremendous emphasis on the study of aggression and selfishness. Fear and aggression have been shown to be very powerful, biologically based characteristics, and in applying these discoveries to man the question is raised whether such aggression can be "controlled."

For example, in Ledyard Stebbins's excellent book there is a section entitled "Human Aggression and Social Harmony," the culminating section on sociobiology. It speaks almost entirely about aggression and how it may be controlled. And Edward O. Wilson, the founder of sociobiology, in a chapter on altruism in his book *On Human Nature*, says: "Human beings appear to be sufficiently selfish and calculating to be capable of indefinitely greater harmony and social homeostasis. True selfishness, if obedient to the other constraints of mammalian biology, is the key to a more nearly perfect social contract."[10]

In this chapter, we have seen that there is indeed a genetic basis for mutual cooperation leading to the building of societies. This inherent tendency of individual members of a species to help other members which they recognize as kin is known as kin selection. We have seen how, in the higher primates and humans, genetically inherited structures provide the energetic basis for behavior which is then molded by culturally transmitted or learned patterns in a very complex interaction. We may wonder whether the inherent natural sense of cooperation implied in kin selection, together with the capacity for learning, may be a significant factor in the development of harmony in human society.

I have tried to indicate that the same biological observations on which statements such as that of Wilson are based—kin selection, the genetic and cultural transmission of behavior, the survival value of consciousness as empathetic imagination—need not be interpreted as supporting the primacy of selfishness and the struggle for survival. While recognizing the powerful grip that selfishness and the struggle for survival do have, I am suggesting that the factors of kin selection and so on can also be the biological basis for the development of friendship and harmony in humans at a more fundamental level than merely the external control of aggression. They can be the basis for overcoming aggression at its source in the individual behavior of human beings.

Many people believe, rather vaguely, that they could change their behavior if they really wanted to. And we exhort each other individually or nation by nation to behave ourselves and try to be better, as if human nature were our own design and any negativities could be transcended merely with the wish. Upholders of various ethical systems have constantly challenged us to try to follow certain guidelines of behavior. Throughout all these systems the message is really rather similar: try to regard others as you regard yourself. However, the practical *means* to do this has rarely been provided. We have been told, simply, to believe in something. Yet the scholar who writes papers or the poet who writes poems on the beauty of human nature, and then retires home to be thoroughly mean to his family, is almost a cliché. And some of our best and most liberal leaders argue that the private life is irrelevant to the public statement. And now we have, for several generations, told each other to believe that human nature *is* after all fundamentally selfish and that we should accept this and try to make the best of it.

The main theme of this book is, of course, that we *can* uplift our own lives and those of others and discover our unconditioned nature which brings dignity and goodness. Such possibilities are based on a knowledge of human nature, as it is in all its details, negative and positive, as we have begun to see in this chapter. They cannot be based merely on *ideas* about human nature whether those ideas are "spiritual" or "practical." Although human nature does have a very powerful self-centered and self-serving, as well as self-denying aspect, this is not its basis. I am suggesting that there is fundamental awareness which goes beyond this deep-rooted sense of the importance of self. When we act from this fundamental level, we do indeed begin to regard others equally with ourself. However, this is not merely a philosophy or an ethical system to believe in. There is a specific practical means by which we may if we wish actually see our own nature, its selfishness as well as its basic goodness. This is the practice of sitting meditation.

Based on the insight gained through the practice of meditation, a very commonly taught contemplative practice in the Buddhist tradition is known as the development of *maitri*. *Maitri*, a Sanskrit word, means friendliness or gentle kindness. The practice of the development of *maitri* is the practice of acknowledging our innate feeling of kindness toward oneself and one's family and gradually extending this to others. It is based on an intrinsic recognition of precisely the factors we have been discussing in this chapter. It is interesting to note that both the words *gentle* and *kind* are related to the idea of family: *kind* is directly derived from the word *kin* and *gentle* has its root in the Latin *gens* meaning family. This is again, perhaps, an acknowledgement of the fact that friendliness begins with the family.

The *maitri* practice begins with the recognition in sitting meditation of a sense of warmth and friendliness toward oneself, springing fundamentally from our unconditioned nature. This feeling of friendliness is then gradually extended, in the contemplative practice, first to one's immediate family, one's parents and brothers and sisters. Beyond this a feeling of friendliness is radiated to one's close friends, to one's colleagues and acquaintances and beyond these to all humans and finally to all sentient beings.

It is the uniquely human possibility that our awareness can be trained. We can become aware of our biological and cultural inheritance as they directly affect our behavior moment by moment. This awareness then gradually transforms that behavior by grounding it in a vaster perspective.

In this chapter, we have seen that forms of organisms arise not only from the principle of struggle, but also from the principle of cooperation or kinship. Thus families and societies of organisms arise which can themselves be considered living units. In the next chapter, we will see that the notion of "living organism" has expanded far beyond the simple idea of a "thing" that is living to encompass patterns of organization on a vast scale.

8   Self-Organization and the Pattern of Life

In Chapter 6 we saw that while neo-Darwinism has unquestionably provided tremendous understanding of the evolutionary process, nevertheless natural selection through struggle for survival is by no means fully adequate as an explanatory principle. In particular, we saw that the sudden appearance of new organs is extremely difficult to explain on neo-Darwinist principles. When we look at the evolutionary process not in terms of the details of how particular species may have evolved but in terms of the large-scale changes of the population of species altogether, we find some features which again are inexplicable on orthodox principles.

An Inexhaustible Variety of Forms

It has been estimated that at the present time there are 1.6 billion different living species and that the total number of species that ever existed is between 1.6 and 16 billion. Within each species there are tremendous variations of individual expressions of genetic traits. For example, a plant which usually grows in the California coastal ranges was dug up and replanted in four very different environments. The resulting plants were strikingly different, so that botanists finding them might very well have classified them as different species. In contrast, there are many cases where genetically different organisms appear identical. Of each of the 1.6 billion species, there can be as many as billions of embodiments in existence at any time with as many different forms. Thus, the entire mass of living organisms has an aspect of continuity to it, an aspect of process and of flow, the underlying energetic process of which individual forms are the discontinuous expression.

The distribution of this almost inexhaustible variety of species at any time—for example, the present era—and its variation over time cannot be explained only in terms of "struggle for survival." We might ask, in general, why was such a vast number of species necessary, if only the "fittest" have survived? Why did not the few best fitted take hold and continue, perhaps changing from time to time as geological and meteorological conditions changed? On the other hand, we might ask, given that so many species *did* evolve, why have there been so few major divisions (phyla), estimated at about thirty?

However, our questions can be much more specific than these, and they have been carefully documented in *The Great Evolution Mystery*, by Gordon Rattray Taylor, who was for many years chief science adviser to BBC Television.[1] There is the question of the diversification of species: J. C. Willis, working in Ceylon, investigated the number of species into which various genera of plants diversified. He found that the numbers followed a quite regular statistical distribution, some genera containing only one species, others as many as thirty. He found many instances in which, of two very closely related species, one was widespread throughout Asia and Africa, while another was known only in Ceylon. "Survival of the fittest" can explain neither of these observations.

Next, we have the phenomenon of "explosive radiation": phyla have diversified very little for perhaps millions of years, whereupon they have suddenly radiated into numerous different forms. Mammals and reptiles provide examples of this. On the other hand, there have been phyla which have continued to evolve very slowly, the bivalves such as oysters and mussels being examples. Other species have not evolved at all, even when closely similar species *have* evolved in the same environment. The opossum is an example of this, sometimes being called a "living fossil." The pace of evolutionary change appears to be very uneven, which suggests that there is more than a single major factor at work.

The next group of phenomena Taylor describes are the clear evolutionary trends both in the entire organism and in specialized organs. For example, we often find that, in an evolutionary series, there is a gradual increase in size of the organism, sometimes followed by a decrease. The ancestors of the horse, the camel, and the llama all seem to have been roughly the size of a present-day hare. Marsupials, which started out the size of a rat, evolved to the size of a rhinoceros and declined again. Individual organs have often developed to such a degree that it appears absurd to attribute greater "fitness" to them. One example is the saber-toothed tiger, whose canine teeth evolved to be so large that it could not close its mouth. Another example is the giant antlers of the Irish elk. Walter Modell, a specialist in antlers, has said, "The antler is a strange and uneconomic experiment, extremely costly to its possessor in many ways."[2]

Another phenomenon hard to explain solely on the basis of superior "fitness" is parallel evolution. Orthodox evolutionists sometimes argue that the appearance of similar forms in different geographical regions is evidence *for* selection of the fittest. However, the contrary could just as well be the case. As Taylor remarks, "Why did Australia and South America, which are not particularly alike environmentally, both contain marsupials? And why did the Old World, which includes environments resembling both of those continents, *not* produce any marsupials?" Another example is that of frogs, in which changes of the secondary characteristics of flattening of the skull and reduction of cartilage bone occurred at different rates in different evolutionary lines. Eventually, the lines became quite unlike, although all were subject to the same environmental pressure. D. M. S. Watson, who made these observations, remarks, "It seems in every way likely that this flattening is brought about by internal factors not directly influenced in any way by the environment."[3] We might argue that such secondary characteristics may be relatively unimportant and therefore not significant as counterexamples to survival of the fittest. Yet it is just such secondary characteristics—number of bristles on a fly, coloration of moths—that have been the main source of support for orthodox argument. For the orthodox view, *all* characteristics should arise via direct environmental selections.

Finally, we have preadaptation, the appearance of structures which are not yet needed for "fitting" the environment, but which will come in useful millions of years down the road. For example, feathers seem to have appeared, unnecessarily, long before flight. The orthodox argument is that the feathers were for warmth, yet hair which had been around for a long while would have served this function just as well.

On all of this, Gordon Rattray Taylor comments, "Far from being a struggle to survive, it looks more like a glorious romp." This light-hearted remark gets right to the core of the matter. The profusion and richness of forms; the trends or apparent "foresight" of natural evolutionary processes; the changes of form of species, independent of the environment according to some internal pressure, all testify to the richness, creativity, and, in a certain sense, intelligence of nature, the presence in nature of mutual cooperation, and perhaps even "mental process." As even one of the great founders of neo-Darwinism, Ledyard Stebbins, says, "Contemporary evolutionists often describe evolution as a series of games—rather than a series of purposeful competitions—that are played because they are inevitable."[4] Explanations of all these phenomena on the basis of "struggle for survival," while thought "plausible" by their proponents, are considered "contorted and highly speculative arguments in an attempt to justify the status quo" by Taylor and many professional biologists.

The picture that has been presented seems to point in the direction of some kind of dynamic organizing principles in the structure of organisms themselves as these organisms manifest themselves as members of species changing over time. Such principles do not have to be supernatural, but they do demand a broader approach to evolution. A broader approach involves recognizing the processes, the patterns, and the organizing principles which give rise to particular forms.

We usually think of our world as being filled with essentially unchanging objects. The changes that we see happening to these objects we tend to regard as of secondary importance. For example, we might have a pet dog, Rover. Over the years Rover gets sick, grows thinner or fatter, shows gray hairs, and so on, but we still think of him as the same dog, the same Rover. As we showed in Chapter 3, this fundamental assumption about our world comes in part from our use of language and our taking for granted that things are like words. We will begin to discover, however, that far from being filled with essentially unchanging objects, our world might be viewed as interlocking processes of continual change in which there is natural order which is continually increasing as well as breaking down.

That we perceive cetain of these processes as relatively unchanging objects—trees, cows, rocks, clouds, galaxies—is also due in part to our scale of perception. If the time scale of our perception were extended so that we saw an hour as one moment, then clouds would seem to be bubbling, seething process. If our time scale were further extended so that a year seemed as one moment, then a tree would be seen as a dynamic process. If our time scale were even further extended so that a million years seemed like one moment, then a galaxy would likewise seem to be in rapid process. Conversely, if the time scale of our perception were reduced so that we were able to detect one-hundredth of a second as one moment, then we would again begin to see changes that are not now perceivable. This last possibility especially is not merely a philosophical point, since such refinement of perception to finer time intervals is one of the secondary results of the practice of meditation.

This consideration of the relationship between unchanging form and dynamic process applies in particular to our ideas of individual living organisms and of the species they embody. To speak of various kinds of finches and wonder how they got to be that way, or to speak of the analogies in structure between the limbs of a human, an ape, and a bird, and to wonder how one form changes into another form, is still to be caught in the preevolutionary conception of a static world of forms. True, the forms which are manifested now change with time instead of being fixed for all eternity. Still, the emphasis is on the forms themselves. Yet these forms are but temporary cross sections of a continuous dynamic process. The form of the individual from birth to death, or the species from first appearance to final extinction, is never in fact fixed; it is always changing. We think about this process in terms of fixed forms changing into one another, like individual frames of a movie. But this is not how it is in nature. In nature there is only process and continually changing patterns. We could understand process itself as fundamental, the forms being momentary appearances of this process of greater and lesser duration. Instead, we think in terms of fixed forms, and we see change as a continual falling away from or struggling to maintain these forms. Thus we form the idea that forms are continually "decaying" and fail to understand how complexity and order can evolve in face of this decay. We will see that complexity and order in form can in fact arise, or be self-generated from within a system. In order to do this we need to take a small detour into the physics of order and chaos.

Entropy: "The Universe is Running Down"

A friend of mine has a motor car that has traveled 130,000 miles. Every few months something goes wrong, and he has to take it in to get it fixed up. He tells me that at these times he feels an utterly irrational anxiety and depression. He knows that the car will eventually run down and stop forever. But perhaps he has a vague, unspoken hope that this will not really happen, and a vague, unspoken fear that it will. Every little problem is a sign that what he fears has finally happened: that this is the end.

We learn in school that the universe as well as our own bodies is just like that car, or any other machine: it is running down. At the beginning of the century, the inevitable "heat death" of the universe was a topic of general fascination. Eventually all the heat energy in the universe would be uniformly spread throughout it, just as when we pour boiling water into cold water it quickly mixes to make tepid water. Then nothing other than completely random motion would continue for eternity. It is still proclaimed in popular and elementary school science books with a kind of perverted glee: "The universe is running down." The dictum has been incorporated into the fabric of our culture as one more factor in the general gloom about nature: not only is nature fundamentally futile, but it is even sapping our energy and breaking down all that we create. At the same time, we learn that there is in nature an accident, which is life, and that life, through struggle, has evolved to man. Man's rational mind, able to stand outside nature and obtain objective knowledge of it, can therefore witness and control it, and bring a small island of order into the brute chaos.

Yet the inevitable decay of the universe, the accidental appearance of life and man's futile but noble struggle against such decay and ultimate chaos, is a myth which dominates us. It dominates our social ideals, and it dominates us individually. By "myth" I mean a partial truth which has become embedded in our culture as an ultimate principle, and as metaphor has embodied many levels of meaning. This myth has to do with the nature of change and time and in particular with the irreversibility of time. In fact, classical mechanics implies that we could simply run the world backward like a movie, and all the laws of physics would still be obeyed. For example, if you saw a movie of balls rolling and colliding on a billiard table, you could not tell whether the movie was being run forward or backward.

Now consider a simple incident such as baking a cake: you break eggs into a bowl, mix them with other ingredients, and place the batter in the oven. Half an hour later you have a cake. Two things have happened here, the destruction of order when you broke the eggshells, and the creation of new order when you mixed and baked the cake. It is because there is this destruction and creation of order in the world of our experience that time does not seem to be reversible. For example, if you saw a movie in which some playing cards lying on the table rose up into the air to form a card house, you would know that the movie was being run backward. But in the biological world, the *creation* of order also leads to irreversibility in time. For example, you could return a tree to the earth by chopping it down, cutting it into small pieces, and leaving it on the ground to rot. But you could not return it to the earth by reversing the precise sequence in which it was formed: adult tree, to sapling, to seedling, to seed. Likewise you could, eventually, return a chicken to the earth by cooking it and eating it. But you could not send it back along its growth path into a chick and finally an egg. This is not merely because of your human inadequacy but because the process is *essentially* irreversible.

It is the irreversibility of the physical world, together with the mechanical model of the universe, which led physicists at the end of the nineteenth century to the second law of thermodynamics, sometimes called the entropy law, and from this to the idea that "the universe is running down." If you put some sand at one end of a tray and shake it up, the sand will eventually be spread uniformly all over the tray. If you put some laughing gas in a small capsule in one corner of a room and break the capsule, the gas as well as the laughter will eventually spread all through the room. Now, since the universe was conceived to be like an empty, infinite space with bits in it just like the grains of sand, it was naturally thought that these bits also eventually would spread uniformly throughout, resulting in "heat death." To put this more precisely, Rudolf Clausius, who formulated the entropy law in 1865, realized that if a system was completely isolated from its surroundings it would always move from a state of greater order to a state of lesser order. Or to put it the other way round: a system will move from a state of less randomness (chaos) to one of more randomness. As the system changes its state from greater to lesser order, its total energy, which must remain constant since the system is isolated, will change from a more usable to a less usable form. For example, if gas is concentrated at one end of a cylinder, its energy would be usable. We could us it to push a piston. But if we allow the gas to expand, while keeping the cylinder isolated, the energy of the gas will no longer be usable; it will just have raised the temperature of the cylinder a little. The randomness of a system is known as its *entropy*. And it is the tendency of a closed system to move to states of greater *entropy*, thus losing usable energy, that is the entropy law.

Nineteenth-century physicists thought of the entire universe as a closed system, and that is why they thought it was running down. However, to speak of the universe as "closed" or "open" is meaningless. Our question is, can we in fact extend the entropy law as a general principle forming one of the background presuppositions of our life and governing everything we do?

To be more specific, should we take as an ultimate principle that every system, our own body/minds, the organization of our lives, our social organizations, will inevitably run down to its most chaotic state unless we maintain it by some external constraints? That is, can order, complexity, and harmony be generated within a system, or can it only be imposed from without? Certainly there seem to be aspects of our experience that contradict such an ultimate principle: the fact of evolution itself, that extremely complicated and sophisticated organisms such as men and women have evolved from primitive single cells; embryology, the fact that a complex individual develops from a single, almost spherical cell; the process of learning, that we can learn and retain complex intellectual, physical, and social skills; the process of sitting meditation, that by paying attention to our thoughts, emotions, and physical processes a sense of insight, energy, and harmony develops; the process of spiritual growth, that our whole life can be a journey of opening and extending our own vision and our capability to help others. Are we to regard these as all simply illusion, accidents in the grand design of steady degeneration? Do we have to invoke some supernatural principle entirely outside of the natural world to understand this appearance of increasing order? Or is it possible that the inevitable decline to chaos is *not* an ultimate principle of nature at all?

Chaos and Order in Open Systems

In the past two decades an entirely different understanding of the entropy law itself has developed, in which the principle figure has been Ilya Prigogine, 1977 Nobel prize winner in chemistry.[5] Prigogine begins by considering open systems, systems in which there *is* exchange of energy and matter with the environment. The strict classical formulation of the entropy law was for closed systems, although it is very difficult to find or even conceive of one existing in nature—perhaps an astronaut floating in space in a spacecraft completely encased in lead would be the closest we could come. Prigogine points out that even some very simple open systems, when energy is applied to them from an external source, enter a stable state which is *not* the most disordered. For example, if a tube containing a mixture of two gases is heated at one end, one of the gases concentrates at one end of the tube and the other concentrates at the other end. This is not the state of highest disorder. As Prigogine succinctly remarks, "This shows that non-equilibrium can be a source of order." This observation initiated his work. Another simple example is a horizontal layer of fluid between two parallel plates heated on one side. Convection currents are set up which should increase the disorder, due to stirring. In fact, a highly ordered pattern of convection cells forms.

The work of Prigogine and his colleagues, then, consists of the detailed study of structures which are in a state neither of maximum stability, nor of maximum disorder, but which are nevertheless stable, and which exchange energy with their environment. He calls these *dissipative* structures, since their stability depends on dissipation of energy from and into the environment. He studies in detail complex chemical reactions and shows that indeed the remarkable patterns of dynamic ordering which emerge in such systems can be understood as dissipative structures. Such dissipative structures occur when the chemical reaction is far from equilibrium, that is, far from the lowest-energy, maximum disorder state.

Prigogine shows, furthermore, that the occurrence of such stable, high-level ordering depends on the global, overall characteristics of the system: the overall size, complexity, distribution of substances in space, and the history of the reaction in time. In other words, it occurs in systems which involve long-range order through which a system acts as a whole. According to Prigogine, "this global behavior greatly modifies the very meaning of space and time. Much of geometry is based on a simple concept of space and time, generally associated with Euclid and Galileo. It is quite remarkable that this simple conception of space and time may be broken by the occurrence of dissipative structures. Once a dissipative structure is formed, the [uniformity] of time as well as space, may be destroyed. We come much closer to Aristotle's 'biological' view of space-time."

The state of a dissipative structure depends on its history. As it develops in time, such a structure reaches certain branch points (known as bifurcations), and which path of development it "chooses" to take at each branch point is not determined by the conditions; it is in that sense random. Thus the growth of a dissipative structure involves both determinism (between branch points) and randomness (at branch points). The dissipative structure passes through each branch point in response to increases of energy input from the environment. As it does so, it enters a state of increased complexity, and increased order. If it does pass through to a new state of order and complexity, then even further, higher-order states become available to it which were not available before. This path from complexity to higher complexity is irreversible, except by sudden collapse in response to extremely high energy input. A system may respond to increased energy input by collapsing back to a more disordered state, as well as by passing through a branch point to greater order.

We can see many examples of just this process in nature: the changes in patterns in a water pipe as we increase the flow rate; the response of a child to a new learning situation; the response of a community to the sudden appearance of a new industry; the response of a society to a new opportunity or threat. Each of these can be met with collapse to a more primitive state, or rising to a higher level of order and energy.

In summary, a dissipative structure is a system exchanging energy with the environment in a dynamic, metastable way, a structure which is self-organizing and self-sustaining and which depends in a unique way on the "choices" in the branch points on the pathway of its development, or its history. This may seem rather an abstract concept, but Prigogine and his team, having developed the mathematics in relation to chemical reaction, go on to apply the principles to biological systems, such as enzyme activity cycles in metabolism; to ecological systems, such as the fluctuations in populations in an ecological niche; to the construction of pillars and arches by termite colonies; and to human social systems. Erich Jantsch, in *The Self-Organizing Universe*, has worked out these possibilities in some detail.[6]

There is perhaps some danger that the tremendous enthusiasm about this view will result in another hand-waving ideology. Nevertheless, Prigogine's work is important in showing us that the occurrence in nature of self-organizing structures of ever-increasing complexity and richness, including intelligence, is not incompatible with the entropy law. In fact, the appearance of such structures follows quite naturally from application of the entropy law to open systems. Furthermore, the appearance of such structures is not at all inherently bound to "struggle" or aggression. It is rather a result of the richness of possibilities.

Darwin sought a mechanism for evolution based on how fixed forms might change, one into the other. Thus, almost inevitably he came up with the idea of natural selection through struggle. However, now we see that we do not need at all to seek a mechanism for change, for the appearance of increasing richness and order. From this point of view, natural selection becomes a sculpting and refining process, rather as a gardener trims a rosebush. Natural selection may be seen as a cosmic gardener, but it is not the source of nourishment, or the basis of complexity and order.

The recognition of the possibility and actual existence of self-organizing exchanging formations (dissipative structures) now gives us a different and unified way of viewing space-time and its "contents." I have used "exchanging" here rather than Prigogine's "dissipative" to emphasize that their inherent nature comes from the fact of their exchange of energy with the environment; they are not fundamentally separate from the environment. And "formations" emphasizes that they are "forms," "appearances," "processes" rather than substances or solidnesses of any kind.

We will see later that elementary particles can be regarded as self-organizing, exchanging formations. At the next level, molecules and especially the key molecule in the chemical basis of life, DNA, are self-organizing, exchanging formations. DNA is self-organizing in that it dynamically maintains its own identity from generation to generation. In fact, Prigogine proposes that the appearance of such molecules, and therefore of life itself, may be due to the possibility of metastable self-organizing states in the primeval soup, the rich ocean of organic, but nonliving, chemicals (amino acids and so on) thought to have existed on earth before life appeared.

The Organization of the Living

The discovery of the DNA-based mechanism of inheritance and protein synthesis had the profound consequence of appearing to provide a material-based definition of the distinction between "living" and "nonliving": a system in which DNA is capable of reproducing itself and directing the manufacture of the organism of which it carries the genes could be called "living." However, this then raises some further questions. There are some organisms which in this view would be "living" under some conditions and "nonliving" under others. For example, some viruses, such as the virus T4 which infects the bacterium *Escherischia coli*, will remain stable and unchanged for many years if they are kept in a bottle of sterilized broth.

However, if the virus after these many years is placed in a bottle containing broth and "living" bacteria, then it infects the bacteria and reproduces itself until there are no bacteria left; that is, the virus becomes "living." Thus, for a virus to be "living," it needs certain components and complex metabolic cycles in its environment which it itself does not carry. This suggests that the criteria for "living" are not based in the material—the chemical composition of the DNA. Rather, the criteria are in the dynamic patterns of structure and interaction. Considerations such as these have led some of the more thoughtful and adventurous biologists to come to a definition of life similar to the following, that of Feinberg and Shapiro: "Life is fundamentally the activity of a biosphere. A biosphere is a highly ordered system of matter and energy characterized by complex cycles that maintain or gradually increase the order of the system through an exchange of energy with its environment."[7]

Single-celled organisms, complex organisms, biological societies, cultural groups, and even the earth's biosphere itself may be regarded as self-organizing exchanging formations. The earth's biosphere has been shown recently, by James Lovelock, to behave in many ways as a "living" organism in the way it sustains its integrity. Since the biosphere receives a constant supply of energy from the sun and from cosmic radiation, and is constantly undergoing highly complex chemical and biochemical reactions, it is precisely the kind of system to which Prigogine's theories apply. This hypothesis Lovelock calls the Gaia hypothesis after the Greek earth goddess, Gaia. He presents the case for Gaia through a detailed study of the composition of the atmosphere, which is statistically so highly improbable that there must be a global mechanism to sustain it. He then investigates and elaborates this mechanism through study of the cyclic interaction between atmosphere, oceans, and biosphere. The Gaia hypothesis is, says Lovelock, "that the entire range of living matter on Earth, from whales to viruses, and from oaks to algae, could be regarded as constituting a single living entity, capable of manipulating the Earth's atmosphere to suit its overall needs and endowed with faculties and powers far beyond those of its constituent parts."[8]

In their audacious book *Life Beyond Earth,* Feinberg and Shapiro make use of this definition to suggest that the search for extraterrestrial life is quite misplaced if we are merely looking for chemical or biological forms that are associated with life on earth. They argue that it is the patterns of organization and the processes of change that would be indications of life. They then suggest that such patterns and processes might be found in balloonlike organisms in the atmosphere of Jupiter, plasma life in the interior of stars, radiant life in interstellar gas clouds, solid-hydrogen life on very cold planet surfaces and other extraordinary places.

A way of looking at the organization of the living, based on considerations of internal structuring, was introduced by Humberto Maturana and Francisco Varela in the early 1970s.[9] Maturana first makes the seemingly obvious statement that "everything that is said is said by someone." Obvious though it may seem, this statement is very important to our understanding of what constitutes a scientific description. It goes back to our discussions of language and of scientific method. Its truth is continually being forgotten not only by conventional scientists but even by those striving to formulate a "new paradigm." Essentially, what it is saying is that we should always try to bear in mind, in each of our descriptions, which elements of that description are elements of the explanatory conceptual system itself, and which are introduced only because of the need of the observer to communicate in a certain way.

Applying this clarifying view to the various definitions of "the living," Maturana and Varela show that most such definitions do in fact confuse those elements which the observer has chosen to emphasize in his description with the elements needed in a description of the organism taken as an autonomous whole. In particular, as Varela points out, the great developments in molecular biology have been the elucidation of the molecular mechanism of protein synthesis and gene reproduction, which has led biologists (and, following them, psychologists, sociologists, and academic and popular science writers generally) to take as the "key to life" the capacity of genetic material to reproduce and preserve itself generation after generation, through encoding in the DNA. This means that all the properties of the individual unit as a coherent, cooperative whole (the functioning cell, for example) are now said to be embodied in a single molecule, DNA, which now contains an abstract, encoded description of the generative goal of the cell. But the abstract information contained in DNA is not the same as the living cell itself. In order to describe the living unit, a description of the components of the whole cell and their interactions is needed—merely describing the information contained in the DNA will not suffice.[10]

Maturana and Varela then define a living system as a system organized as a network of processes of production of components. These components themselves, by their interaction and transformation, continually regenerate and realize the network of processes that produce them. Furthermore, this network and its components form the system as a unity in space and time. Thus, this is essentially a self-producing and self-sustaining unit, and for this reason Maturana and Varela have coined the term *autopoiesis* to name it: from the Greek *autos* – self and *poiein* – to produce. This is clearly a particular type of dissipative structure, or exchanging formation, in which there is a particular emphasis on the self-defining and self-sustaining unity of the structure as it undergoes changes in its environment. This, then, for Maturana and Varela is the definition of a living system, and from this they are able to build a description of the interactions between such systems and even of the introduction of symbolic communication. Varela, based on his definition of the living as autopoietic system, suggests that we should look at evolution from the point of view of the organism, rather than that of an "objective" observer outside of it watching how it "adapts." When we do this, Varela says, the important point becomes what forms are allowed by the flexibility of the internal structure of the organism. The environment then simply places some constraints within which these allowed forms must stay.

Whereas from the natural selection viewpoint only those forms survive which are *allowed*, from the autopoietic viewpoint all those forms which are *not forbidden* flourish. As Varela says, "The main point here is the questioning of some kind of optimization [the fittest] as the main guideline for evolutionary transformation."[11] This simple but profound change of viewpoint opens up altogether fresh avenues for thinking about environment-organism interaction in evolution, as well as about purely internal forces of change. We can think of the continually changing environment continually opening up further possible habitats for species to evolve *into* through their internal pressures, their "curiosity," and their vast richness of possibilities.

Patterns of Evolutionary and Mental Process

Finally, these vastly broadened views of "living systems" can be extended to include mental processes. In discussing the general characteristics of complex systems, Gregory Bateson proposes six criteria for the appearance of mental process. I will discuss these criteria in detail in Chapter 13. Bateson argues that the patterns of the process we call evolution are precisely analogous to the patterns of the process we call thought. Therefore, since these patterns are all we know, we can say that evolution, as we know it, has the characteristics of mental process: hence the title of Bateson's book, *Mind and Nature: Necessary Unity*.[12] These processes exhibit an alternation between the random and the selected, between form and process, between continuity and discontinuity. They are stochastic or "random walk" processes in which periods of continuity and causal unfolding are punctuated by random changes which are apparently acausal at a certain level of order. A stream of thought continues for a while and then suddenly changes, apparently without a cause, although a selection and reinforcement process is taking place at nonconscious levels. A species continues to evolve, changing according to apparently random genetic factors coming from the interaction of individuals with the environment. We have seen the intimate relation between these two systems in the discussion of the Baldwin effect in Chapter 6. We saw that the same language, the language of behavior, may well be appropriate for both.

Speaking of criticisms that Darwin left no place in *The Origin of Species* for a mind separate from and directing nature, Bateson says, "It turns out however that such critics were precisely wrong in their choice of the correction they would apply to Darwinian theory. We would correct the nineteenth century thinkers, not by adding a non-stochastic mind to the evolutionary process, but by proposing that thought and evolution are alike in a shared stochastic. Both are mental processes in terms of the criteria offered above.

"We face then two great stochastic systems that are partly in interaction and partly isolated from each other. One system is within the individual and is called *learning;* the other is immanent in heredity and in population and is called *evolution.* One is a matter of the single lifetime; the other is a matter of multiple generations of many individuals.

"These two stochastic systems, working at different levels of logical typing (order), fit together into a single on-going biosphere that could not endure if either somatic or genetic change were fundamentally different from what it is.

"The *unity* of the combined system is *necessary.*"

It is important to understand that this has nothing to do with identities between mind-stuff and nature-stuff (or matter), it has nothing to do with substances of any kind. It is saying that all we can know are processes, that mental processes have a certain character, and that evolution exhibits that character. In speaking this way, Bateson has moved the discussion to an altogether different level, and perhaps a more genuine one. Instead of speaking of evolution as some objective fact about real things, he is speaking of it as patterns that we notice in what we know when we separate ourselves out as observers of the process of which we are in fact a part. It brings us back to Maturana's simple but powerful observation that "everything that is said is said by someone."

So perhaps we can say that whatever that great process we call evolution is "working out," it is working it out according to the same processes by which we "work out" our thoughts. To just the same extent that we believe ourselves to participate in our body in "living" "mental" processes, so does the entire biomass appear to participate in such processes, at the level of populations.

In this chapter we have seen that when we take a broad view of the evolutionary process it appears to be a process of the continuity of change of inherent possibilities of complexity, order, and mind. In the next chapter we will begin to look at the context of these patterns of self-organization: namely, space, time, and matter. We will again find far greater richness and depth to this world than we have, in the past few centuries, allowed ourselves to dream of or to experience.

9 Space, Time

Behind an outcrop of rocks on the hillside are pines and spruce, the
power of wind and fire embodied in their curves. Below, in front, a
path sweeps to the right, disappears into a cluster of aspens, and
reappears in the distance winding to the left and up the hill on the
opposite side of the valley. Silhouetted against the sky, an outcrop of
rocks appears in a striking likeness of Marpa, the great translator who
took Buddhism from India to Tibet in the eleventh century. The valley
gently slopes down to a wide stream bed which is thick with various
dark green bushes and aspens. One large aspen standing way above
the others has lost all its leaves. Its black spiky branches form a crazy
framework for the hill behind, and are a harsh reminder of the winter
lurking nearby. Tall grasses of the meadow on the side of the valley are
rippled with waves and swirls of dark and light browns, yellows, and
fading greens. Sky and earth seem to sweep in from left and right in
open, gracious curves. The flat stump of a tree that has been cut for
firewood, a reminder of the power, potentiality, and wakefulness of
humankind, adds its own dignity. Altogether, there is a sense of
harmony, of circularity, and of richness.

Relativity of Space and Time

Newton proposed a space which is absolutely unmoving and
noninteracting with anything in it. Descartes defined a coordinate
system, the Cartesian coordinates, so that any point in space could be
defined by its position in relation to three straight lines: front-back,
left-right, and up-down. Renaissance painters learned how to
construct their images according to Descartes's vision. Euclidean
geometry, which we all learn in school, teaches us that the shortest
distance between two points is a straight line. Thus we learn to take an
empty, straight world that stretches indefinitely in each direction as
the background to our lives.

Within our straight, empty, unmoving space, things move. The way we can normally tell they are moving is because something else remains unmoving. We can tell a cloud is moving across the sky when it passes in front of the sun or the moon. Sometimes this is the only way we can tell. If we want to know which way a storm cloud is moving, we have to try to watch it in relation to the sun or the mountain. We cannot tell whether or not a cloud in the clear blue sky is moving. If we are sitting in a plane, we can tell when it has started moving by looking out at the walkway. But sometimes a strange illusion occurs when we look out at another plane moving along the runway. For a moment we think it is we who are moving, although our body sensations don't correspond; then we realize it is the *other* plane that has started. Galileo himself pointed out that if you were in a boat on a calm ocean with no land in sight, you would not be able to tell whether you were moving. While Newton presumed that there must be *some* absolute, unmoving reference system, which was his absolute space, Galileo, a generation *before* Newton, did not.

One more thing about the way we think about movement in our ordinary world: speeds are simply added or subtracted. For example, if I am standing on a railway platform and you walk by at three mph, your speed relative to me is three mph. If I then start walking alongside you at three mph, your relative speed is zero. If you are walking at three mph on a train in the direction it is moving, and it is going through the station at ten mph, then I will measure your speed as you go past me at thirteen mph. If you are walking in the opposite direction to the train, I will measure your speed, relative to myself, as seven mph. Of course, in both cases, you will measure your own speed, relative to the floor of the train, as three mph.

When Albert Einstein was sixteen years old, he thought about these things particularly in relation to light. He asked himself the question: "How would the world appear if I were sitting on a light wave?" Another way of asking this is, if you were in an accelerating spaceship and looking at your image in a mirror, what would happen as the spaceship approached the speed of light? Well, if light obeyed the same simple rules of addition and subtraction of speeds, since you would now be moving along at the same speed as the light reflected off the mirror, the light would not be moving relative to you. Therefore, the image from the mirror would never reach you: when the spaceship reached the speed of light, the image would just disappear.

Einstein felt that arguments such as this, put of course in more rigorous technical terms, showed that something was seriously amiss in the physics of his day. When experiments were done to try to determine the difference in the speed of light relative to the earth in the direction the earth is moving and in a direction at right angles, the answer was that there is no difference. This was the Michelson-Morley experiment, and the negative result was one of the "two small dark clouds on the horizon" of Lord Kelvin. The two measurements should have differed, if Maxwell's equation and the classical addition rule for relative speeds were true. The fact that they did not differ indicated something very much wrong in the classical picture. Realizing that either Maxwell's equation or the classical addition rule had to go, Einstein took the daring step of retaining the former and casting out the latter.[1]

In 1905 Einstein simply proposed that, indeed, no matter how fast you are moving, light will always appear to you to travel at the same speed. So if, for example, I were on the earth, and you were moving past the earth away from the sun at 9/10 the speed of light, we would still both measure the speed of the light from the sun to be the same. According to the simple addition rule, you would have been moving relative to the light at 9/10 its speed, and therefore relative to you it would seem to have been moving 1/10 as fast. Einstein proposed that the speed of light is constant in every uniformly moving reference frame and that, furthermore, nothing can travel faster than the speed of light. He then went on to consider how we actually define the distance between two points and the interval between two lines. We do this with actual physical measuring rods and clocks. Now, if two observers are moving relative to each other, and they wish to check that their standard measuring rods are the same length and their clocks are synchronized, the only way they can do this is by sending messages to each other. These messages cannot be faster than the speed of light, that is, they cannot be instantaneous. When we work through the details of what this implies, we find that a pure distance measurement, or pure time measurement, of one observer, when expressed in terms of the measurements of the other observer, involves both his distance and his time measurements. It also would depend on his relative speed. So that if you were in a laboratory going at speeds close to that of light relative to another observer, it would appear to him that your measuring rods had contracted, your clocks slowed, and your mass increased. For example, if you were traveling at 99/100 the speed of light, your measuring rod would shrink approximately to one-seventh of its original length, your clocks would be seven times slower, and your mass seven times greater. These effects are observable, of course, only at very high speeds, but they have been observed many times in the laboratory when elementary particles travel close to the speed of light. If you were able to travel at the speed of light, you would have zero thickness, infinite mass, and time would have stopped for you. The answer to Einstein's boyhood question now becomes: if you were able to ride on a light wave, time would stop, you would forever be in the same moment. So the speed of light puts an absolute limit on how fast a signal can be sent. Because of this, as we shall see later, even the definition of past and future becomes different for observers moving relative to each other.

So space and time are now no longer the absolute and distinct qualities of Newton. They are relative to the particular observer, and they are interconnected. Time now becomes a fourth dimension on an almost equal footing with the three dimensions of space. It is more in keeping with relativity physics to think not of an event's taking place at a point in space at some time but to think of it at a single moment in space-time. What we had thought of as an object existing in space as time went by more appropriately would be thought of as an unchanging tube extended in four-dimensional space-time.

In order to help him present space-time to his imagination, I invite the reader to stand or visualize himself standing at a square table in front of a window, with the window on the opposite side of the table from him. Suppose now that there is a one-dimensional being—let us call him Unidem—that can move only along a line in space. This could be the line between you and the window. According to our classical picture of separate space and time, this being is free to move backward and forward between you and the window, along his space line. Now you, representing time, walk along the side of the table parallel to the window at a steady pace, and his space line moves with you. For Unidem this is like feeling his world (his back-and-forth line) moving in time. However, you see his life as an irregular line across the table, i.e., in two dimensions of space-time. This line would be Unidem's version of our four-dimensional space-time tubes. (See Figure 1.)

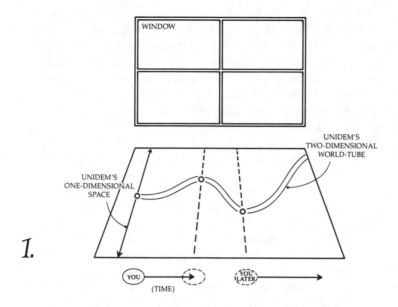

What is "Now"?

We usually preconceive time, as well as space, as a straight line. Between now and nine o'clock tomorrow morning, there is only one time and there is only one way of getting there. If you think of traveling between two towns, there are many ways of getting there: you can take various county roads and detours, or the highway. But there is only one shortest way, and that is straight, perhaps by helicopter. And this is how we think of time: my "now" is the same "now" as everyone else's, and it is the same now as that of the sun and the distant stars. And the interval between this "now" and tomorrow morning's "now" is also the same for me, for you, and for the sun and the stars. The only way to change the quality of "now" is to fill it differently. "Now" has no inherent quality other than this: a container for our activities.

Let us look again at the theories of relativity to see what they might tell us concerning time. As we saw, clocks in a moving reference frame appear to slow down relative to a stationary observer. This can be, and has been, verified in the laboratory by measuring the average lifetimes of particles, for example mu-mesons, traveling close to the speed of light in a cosmic ray shower. Their lifetimes are appreciably longer than the lifetimes of mu-mesons generated in the laboratory and traveling at slow speeds. This is because, while to a mu-meson in the cosmic shower its own clock goes at the same speed as the earthbound mu-meson, to you it appears to go at a slower speed. Therefore, to you, it seems to take longer to decay.

According to Einstein's later, general theory of relativity, clocks are also slowed down by a gravitational field on an accelerating frame of reference. This also has been tested. An atomic clock was flown around the world in a jet liner, and it did indeed slow down a minute fraction of a second relative to a similar earthbound clock, because of the acceleration it went through. This effect gives rise to the famous twin paradox. Suppose you have a twin brother and you are both twenty years old. Suppose in January you get in a spaceship with a clock and accelerate away from the earth at the same acceleration as that due to gravity on earth. After a month you decelerate for another month, come to a stop, and return to earth in the same way. You, according to your clock, have been traveling for just four months—it is May. However, the brother that greets you as you step out of your spaceship will now be sixty years old. I just spoke of our idea that there is only one way to get from now to tomorrow morning, or from January 1 to May 1. Now we see that these two brothers have taken very different routes to get between two points in space-time. While you spent four months in your spaceship, your brother could have married, had children, sent them to college, and retired, all between the same two events in space-time: the event when you said good-bye to your brother and the event, now, when you say hello again four months later (or was it sixty years?).[2]

Furthermore, the idea of universal simultaneity, a universal "now," is no longer acceptable. We can no longer say that things are happening at the same time everywhere in the universe. This follows very simply from the finite speed of light, the fact that this speed is the same relative to any moving reference system, and the fact that in order to determine the simultaneity of two events we would have to send a message between them.

Suppose two long spaceships are traveling relative to each other in opposite directions at constant speed. At the midpoint of one spaceship is an observer, Tom, and at the midpoint of the other is his friend, observer Mary. Since they have no other reference point, Tom and Mary each think that his or her own spaceship is at rest while the other's passes it. At the moment Tom and Mary are opposite each other, a light which is between them in space flashes and can be seen throughout both spaceships. Tom and Mary later meet up and compare notes on the times that the flash was seen at the two ends of both spaceships. If we bear in mind that as far as Mary is concerned, she is not moving, but Tom moves past her, Mary's report will be as follows: "Because I was an equal distance from each end of my spaceship and light travels at constant speed, the two flashes reached the ends of my ship simultaneously. However, by the time light had traveled to the ends of your spaceship it had moved on, so one end of your spaceship had got closer to me and the other further away. Therefore, light took longer to reach one end of your ship than to reach the other. And this is precisely what I saw." Tom, who regarded himself as stationary while Mary moved, will report the opposite: that the two flashes of light reached the ends of *his* spaceship, but *not* Mary's simultaneously. Tom and Mary cannot agree on whether or not two events were simultaneous, i.e., happened at the same "now." The only way to settle the issue would be if we could say who, Tom or Mary, was "really" at rest (i.e., relative to any absolute frame of reference). And according to relativity theory, we cannot do this.

There are still, according to relativity, certain regions in space-time we could call past, present, and future, and therefore there can still be causality. That is, the universe can still be divided into those events which can be connected by a cause-and-effect relationship, which would define the past and the future, and those events which cannot be connected in this way, which defines the present. But precisely how each observer divides the universe into these three regions, past, present, and future, depends on his state of motion. There is no longer any reason to suppose that there is, somewhere in the universe, an absolute clock ticking away the absolute time.

The reader might take a moment to reflect on this. So much of our life is based on the assumption that while things are happening at this moment in my world centered on my body, other things are happening at other places at the same time. Perhaps you sometimes like to remember a loved one in Australia and what he might be doing "now." Which "now" is that? Although, of course, the practical effect at such small distances and speeds (the relative speed of opposite points on the earth) may seem to be negligible, we are talking here about a deep assumption about how things "really" are which conditions our perception. Such ideas have been presented in popular books on relativity for fifty years, but how much do we take these things *personally*? To what extent do we merely say "wow" rather than let them be a knife cutting through the powerful preconceptions that limit our perception by causing us to view the world in such highly abstract terms.

This again raises the question "What is now?" If our four-dimensional tube is a more "real" view of our own life span, then why are we not aware of it in this way? Why do we not see our whole life as existing as one block? We would, of course, have to be out of "time" in order to do this. In the case of Unidem, you are in a third dimension outside of his two and looking down on him. There is no way within physics as we now know it that we can tell whether there is a fifth or higher dimension in relation to our normal four. Nor do we usually experience such a dimension. However, this does not alter the question as to why, if the relativity view is correct, we feel ourselves moving along our life tube "in time" rather than see the whole thing as one. The implication is that in order to bring the relativity picture into correspondence with our experience of reality, something else must be added to our experience, something having to do with "now." Einstein himself thought that indeed our feeling of past, present, and future is a subjective illusion. He wrote to the widow of his good friend Michele Besso, "Michele has preceded me a little in leaving this strange world. This is not important. For us who are convinced physicists, the distinction between past, present, and future is only an illusion, however persistent."[3] However, this viewpoint comes, in part, from thinking of ourselves as a mathematical tube in the space-time of relativity. When we recognize that we are complex organizers carrying on continual energy exchange with our complex environment, a very different picture emerges: then a very different sense of time emerges, according to Ilya Prigogine—a sense of time in which change is irreversible and is punctuated by transitions to increasing complexity and order.[4]

Auspicious Coincidence and the Fullness of Time

Now, finally, before we leave the subject of time altogether, we will discuss another facet of time which has been noticed and discussed for centuries, that of meaningful coincidences. Perhaps of all events pointing beyond the linear-continuous facet of time, this is the one we are all most familiar with. We should distinguish between simultaneous events and meaningful coincidences. If you go to the local store and meet your neighbor, this is merely a simultaneous meeting. If, however, you go to the store and meet a friend whom you have not seen for five years and did not know was in town, but whom you were thinking about that very morning because you rather urgently needed to get some information that only he had, then this is a meaningful coincidence. Michael Shallis quotes an incident reported by Dame Rebecca West, who was searching for an incident in the Nuremberg trials: "I looked up the trials in the library and was horrified to find they were published in a form almost useless to the researchers. They are abstracts and are catalogued under arbitrary headings. After hours of search I went along the line of shelves to an assistant librarian and said: I can't find it, there's no clue, it may be in any one of these volumes. (There are shelves of them). I put my hand on one volume and took it out and carelessly looked at it, and it was not only the right volume, but I had opened it at the right page."[5]

Just as with precognition, there are many many well-documented reports of such coincidences, and we have no more reason to doubt their veracity than we have to doubt the veracity of a scientist's report of his work. Unlike precognition, however, they occur quite frequently in the lives of many people, and often provide a sense of direction for that period of one's life. In fact, as one trains one's attention to be more sensitive to the immediacy of one's life, it occasionally seems to be experienced as a series of such coincidences. While there is much more general willingness to acknowledge that such coincidences happen than is the case with precognition, the issue here, of course, is whether they are *meaningful*.

The key point is the lack of apparent causality in such events, and it is this acausality which causes most "normal" scientists to reject them. We might simply include acausality as one of the facets to be taken into account in any larger understanding of time, as Shallis suggests. We could notice that the experience of both precognition and coincidence has a discrete, momentary, durational quality. Both the event that is precognized, such as the Aberfan disaster, and the moment of cognizing it, the dream, are felt as complete moments of a certain duration, discontinuous from other moments. Likewise, the moment of experience of a coincidence is felt in this way. Again, as our attention is trained we might notice that our lives can be experienced in this way, discontinuously, moment by moment. Thus we should perhaps add to the collection of facets of time, the momentary, discontinuous aspect.

As Shallis says, "Coincidence as a temporal phenomenon is a wonderful example of time's duality. If the concept of the dual nature of light as both wave and particle is acceptable . . . then coincidences should help in the no more strange idea that time displays the duality of the connected, linear causal side of its nature and its acausal, interconnected aspect. The experience of time through coincidence points to a much more complex, much more bewildering and awesome aspect of nature that was overlooked in the more familiar descriptions of the apparently explicable and controllable world given by instructional [normal] science."[6]

This other aspect of time has been recognized in the Western alchemical and hermetic traditions as well as in the Chinese Confucian and Taoist traditions. In the Western tradition, it is connected with the recognition of correspondences between various realms of existence: inanimate, plant, animal, human, and heavenly, or perhaps we would say material, biological, psychological, and spiritual. What appear to be "chance" coincidences when we consider only a narrow realm of experience are recognized to be dependent on multifaceted causes when a larger realm is taken into account. In the Greek tragedies, these larger realms of existence were symbolized by the gods who, while conducting their own affairs independently of man, could also at times enter into and affect the affairs of man—such times being turning points in the lives of those affected, that is, auspicious coincidences.

A very similar situation occurs in Javanese shadow theater. These plays are performed for whole villages usually at night. It is assumed that not only the living human villagers attend, but also ancestors, gods, and all manner of spirits and demons. The latter in fact is the essential audience, while the human audience is, in a sense, just passing through. The language in which the plays are performed incorporates all levels of the history of Javanese language—from Old Javanese and Sanskrit (in which the gods are addressed) to modern Javanese and American. The performance of the play is already, then, regarded as an auspicious meeting of gods and men. This is reflected in the structure of the plays themselves. I will quote at length from anthropologist Alton Becker: "Wayang (shadow play) plots are built primarily around coincidence, a word which we in the West use to explain-away things of no meaning. 'A mere coincidence' cannot, in the West, sustain prolonged scrutiny and analysis. In Wayang theatre coincidence motivates actions. There is no causal reason that Arguna, the frail hero, meets Cakil, a small demon, in the forest as he (or a counterpart) does in each wayang. It is a coincidence, it happens, and because they are who they are they fight and Cakil dies . . . but not forever, he will be killed over and over again in each wayang. This is but one coincidence, one intersection in the interwoven, cyclic actions that inform a wayang plot . . . unmotivated, unresolved, meaningless within a chain of causes and effects, but symbolically very rich."[7]

Within the play, as within the world, there are various worlds and

various times occurring simultaneously and occasionally coinciden-
tally interweaving. "In the coincidence of epistemologies the real
subtlety of the wayang appears. The major epistemologies are 1) that
of demons—the direct sensual epistemology of raw nature, 2) that of
the ancestor heroes—the stratified, feudal epistemology of traditional
Java, 3) that of the ancient gods—a distant cosmological epistemology
of pure power and 4) that of the clowns—a modern pragmatic
epistemology of personal survival." All of these epistemologies coexist
in a single wayang; between each of them there may be a confrontation, a battle; and each exists in a different concept of time, and all the
times are occurring simultaneously. "That is, nature time, ancestor
time, god time, and the present are all equally relevant in an event,
though for each the scope of an event is different."

According to Becker, the wayang is an education in power. The
first line of a Javanese history of the wayang says, "Traditional Shadow
Theatre is a signification of the life of man in this world." The wayang,
then, through its rich symbolism, places the present in the context of
the past and the small human world in its context of the energies and
power of nature and the cosmos. "Wayang teaches men about their
widest, most complete context, and it is itself the most effective way to
learn about the context."

Finally, Becker reports a delightful story that reflects back on our
discussion of the perception of curved and linear space. One of the
functions of the wayang is to subdue madness, demons, disease, and
generally power that has gone amok. In wondering how wayang
controls power, Becker reports: "The closest answer to the question of
how wayang subdues power came to me from a Balinese friend who
answered 'You know, it's like the doors in Bali.' (Note: an entrance in
Bali and traditional Java is backed by a flat wall or screen a few feet
behind the entrance gap in the outer wall, so that one can not go
straight in but must pass right or left. Demons and people possessed or
amok move in straight lines, not in curves like normal human beings).
My friend continued after I look mystified, 'The demons can't get in.
The music and shadow play move round and round and keep the
demons out.' Then he paused and laughed heartily and added, 'As
you might say, demons think in straight lines.'"

Becker concludes, "Shadow theatre like any live art presents a
vision of the world and one's place in it which is whole and hale, where
meaning is possible. The integration of communication (art) is, hence,
as essential to a sane community as clean air, good food and medicine
to cure errors. In all its multiplicity of meaning a well-performed
wayang is a vision of sanity."

In the medieval alchemical tradition, the correspondence was expressed more directly. As Agrippa von Neltershein says, "It is the unanimous consent of all Platonists that, as in the Archetypal world, all things are in all; so also in this corporeal world all things are in all, albeit in different ways according to the receptive nature of each." Or Paracelsus even more directly: "If a man will be a philosopher without going astray, he must lay the foundations of his philosophy by making heaven and earth a microcosm and not be wrong by a hair's breadth. Therefore he who will lay the foundations of medicine must also guard against the slightest error and must make from the microcosm the revolution of heaven and earth, so that the philosopher does not find anything in heaven and earth that he does not find in man and the physician does not find anything in man that he does not find in heaven and earth. And these two differ only in outward form, and yet the form on both sides is understood as pertaining to one thing." As Hippocrates, the great physician, himself said, "There is one common flow, one common breathing, all things are in sympathy. The whole organism and each one of its parts are working in conjunction for the same purpose . . . the great principle extends to the extremest parts and from the extremest part it returns to the great principle, to the one nature, being and not-being."[8] It is through these correspondences between heaven and earth and man that apparently acausal yet meaningful coincidences happen: events that are apparently unconnected from one view become interconnected when a broader viewpoint is taken.

The Chinese conception of the rich, full, and discontinuous facet of time is expressed in the following quotes from Helmut Wilhelm in speaking of the *I Ching*, a Confucian classic. In reading these quotes, we should bear in mind that the more abstract, linear notion of time was certainly known in China at this period and formed the foundation of a quite precise astronomy. Of time, Wilhelm says: "The conception of time in these quotations [from the *I Ching*] is very concrete. Here time is immediately experienced and perceived. It does not represent merely a principle of abstract progression but is fulfilled in each of its segments; an effective agent not only in which reality is enacted but which in turn acts on reality and brings it to completion. Just as space appears to the concrete mind not merely as a schema of extension but as something filled with hills, lakes and plains—in each of its parts open to different possibilities—so time is here taken as something filled, pregnant with possibilities, which vary with its different moments and which, magically as it were, induce and confirm events. Time here is provided with attributes to which events stand in a relation of right or wrong, favorable or unfavorable." And on man's role in relation to time, Wilhelm comments: "A man's relation to time may be taken as a task or as a foreordained destiny. In some situations one can assume a correct or a wrong attitude toward time, while in others one must accept the time as fate. The most advantageous relation to time is naturally that of harmony. In the situations where one is in harmony with time, the maxims of action are a matter of course, or at least they are easy to follow."[9]

Finally, on time, Helmut Wilhelm's father, Richard Wilhelm, the great translator and interpreter of the *I Ching*, has this to say: "To create this kind of harmony, it is essential to find the proper position. And this proper position is in the center. Time, it was stated, is the necessary ingredient that enables us to experience opposites; and experience, in fact, is only possible if contrast is encountered. But we see now also the importance of not being borne along by time alone, for time cannot become a reality unless we have a resting point from which to experience it. As long as we are tossed and torn from moment to moment, reproducing a phantasy of our past in the imagination, or anticipating the future with fear and hope, we are merely objects among many such objects. Mechanically propelled by our fate like all other purely mechanical objects, we are moved here and there by thrusts and counterthrusts. However if we succeed in experiencing time, including its opposite from a central point of view rather than withdrawing from it then the circle will begin to close, we experience time as perpetuity. This consists precisely in time becoming harmonious."[10]

Time, then, has tremendous richness. In addition to its linear and continuous facet, it appears to have a durational, momentary, discontinuous facet. Various aspects of such moments on all levels appear to be interconnected in a meaningful way and to include a sense of the potentialities of future moments. Presumably the depth of interconnectedness and potentiality of each moment is as unfathomably profound as the world we live in.

Curves and Bubbles in Space–Time

Now let us turn to a consideration of the relation between four-dimensional space-time and matter. Ten years after publishing his special theory of relativity, Einstein presented the general theory. This had consequences even more unsettling for our inherited beliefs. He pointed to the equivalence of a gravitational field and an accelerating frame of reference. In the gravitational field of the earth, for example, all objects accelerate toward the center of the earth at the same rate, *g* (32 ft/sec/sec). Now, if you were in a spaceship which was accelerating at the same rate, *g*, you would have no way whatsoever of telling whether you were at rest on the surface of the earth or in such a spaceship. All objects would, if you dropped them, fall to the floor of the spaceship with an acceleration of *g*. You would feel yourself pulled to the ground just as you would on earth, you would be able to play baseball or cricket just as you had trained on earth, and so on. Thus Einstein argued that the presence of a mass in space is exactly equivalent to the presence of accelerating reference frames. This latter, in our language of four dimensions, means *curved* space-time. If Unidem's tabletop world was now a rubber sheet, then everywhere a mass, say a golf ball, was placed on it, the rubber sheet would curve. And whenever another mass, say a ping-pong ball, was rolled near this dip, it would curve toward the golf ball as if it were "attracted" to it, when in fact it would simply be moving along the curves in space-time generated by the golf ball. Four-dimensional space-time behaves analogously to that rubber sheet near all masses. Near very large masses, this curvature is large enough to be detected, as it was for the first time by Eddington in 1917 when he showed that light rays bend as they pass the sun.

In summarizing general relativity, Einstein is said to have remarked that whereas in Newton's universe if matter were removed, space and time would remain, in the universe of relativity if matter were removed, space and time would go as well. Not only is space-time curved near masses, but the entire universe is curved. Thus, instead of thinking of our universe as a rectangular box of unlimited extension, it may be more appropriate to think of it as the four-dimensional equivalent of the surface of a sphere, a hyperbolic tube, or a donut with a "hole" in the middle. If a mass is sufficiently large, it can cause such distortions in the space-time near it that space-time could close entirely around on itself. This would be like placing a mass on Unidem's rubber sheet which is so heavy that the rubber sheet breaks off and surrounds it. Around such an object there would be a space-time barrier known as an event horizon, beyond which it would be impossible to pass since, as we approached closer and closer to it, our speed would become nearer and nearer to the speed of light. Similarly, no matter or radiation, or messages of any kind, from the inside of the hole could get out. Thus, it is completely and forever isolated from our universe—hence the term "black hole"—and can be regarded as a separate universe unto itself. Inside the event horizon, at the central region of the black hole, conditions would be as they were in our universe before the Big Bang, that is, before the laws of physics as we know them began to operate.

It is believed by astronomers that the presence of such objects has been indirectly detected, and it has even been speculated that there may be such a black hole at the center of our own galaxy. It is also speculated that it could happen that the event horizon of such an object could open up so that astronomers would be able to see directly into a region of space-time in which laws of physics altogether different from ours hold. Our own universe has its event horizon beyond which all communication is impossible. To understand this we need to know that it is now thought that space-time itself is expanding, and all the galaxies in it are moving apart from one another. It is as if Unidem's world, rather than being a flat rubber sheet, were the surface of a balloon, with dots painted on it to make the galaxies. As we blow the balloon up, the whole surface gets larger and the dots move away from one another. Now, in our universe, the further apart galaxies are, the faster they move relative to one another. So at a certain distance from our own galaxy (the Milky Way), other galaxies are moving, relative to us, at the speed of light. It is impossible for us even to observe beyond this. This is our own event horizon.

We can consider space-time, or the gravitational field, to be similar to the electromagnetic field or the fields of the elementary particles we will discuss in the next chapter. Then there should be waves in space-time, similar to electromagnetic waves. Such gravity waves are extremely difficult to detect, but physicists are convinced they are there and have been searching to detect them directly for many years. It is believed that these gravity waves have been detected indirectly by their effect on certain astronomical objects.

The other consequence of regarding space-time as a field is that it should also exhibit particlelike effects. So if we could look closely enough at the tiniest intervals of space-time, we would see it beginning to break up, like froth on the surface of the ocean. According to Paul Davies, "Before these ideas came along, a lot of scientists tacitly assumed that space and time were continuous down to an arbitrary scale. Quantum gravity suggests instead that our world canvas not only has texture, but a foam or sponge-like structure, indicating that intervals or duration cannot be infinitely subdivided."[11]

Of course, these bubble effects take place only at unimaginably minute subdivisions of space-time. But they imply that we can no longer take space and time to be continuous. If space and time can be broken up in discrete bits (known as Jiffies) however small, then they can also possibly be discontinuous at large scales. The effect of such new knowledge is to completely undermine the presuppositions of our experience. We cannot any longer take it as an ultimate given that space is uniform, straight, and unbroken, since the latest news is that it is grainy, curved, interacting with time, and dependent on the presence of matter. And in the next chapter we will find that this "matter" itself appears to be various levels of surfaces in a space that, far from empty, is full of seething oceans of energy or virtual particles. Jiffies and virtual particles are of course not the final story. Physicists are working to join these various versions of matter and space-time, via the field picture, into a single view—a unified field theory. It may even be that in order to develop such a theory they will again have to postulate, at a more fundamental level, absolute space and time as a background to Jiffies . . . . Theories within theories within theories— none of them telling the final truth at all.

These are fascinating tales, just as fascinating as those of Newton and Aquinas before him. Just like those, these draw out certain patterns in nature which we can see if we look in a certain way. They, too, illuminate the prejudices and partial views we take so much for granted, and the thought forms we take as ultimate truth. We have seen that only on one relative level, the level of sun, earth, moon, motorcars, and astronauts, is it reasonable to say that space seems noninteracting, uniform, and empty. At scales much smaller or much larger than this, space appears altogether different: full, seething with energy with which it interacts, itself swirling with waves of four-dimensional expansion and contraction and, at the smallest end of the scale, grainy, foamy, discontinuous. The important point is not so much which of these views is more "true" but to see that unexamined belief in either of them conditions our experience very profoundly.

In a very careful and remarkable study, Patrick Heelan has shown that our preconceptions of a straight-line world do profoundly affect our ordinary visual perception.[12] He shows that another mode of perception is available to us, which he calls hyperbolic perception, in which we perceive the world as if along a curved grid rather than a rectangular one. Most people in our scientific culture experience such perception at rare moments of perhaps particularly relaxed gaze, or when shocked by a visual illusion. However, Heelan argues that some painters may have perceived in this way continually. For example, the paintings of Van Gogh may not have been the distortions of extreme astigmatism or a crazed mind. It is possible that they were direct images of the way Van Gogh perceived the world—without the usual, rectangular, preconceptual filter. (See Figure 2.)

2.

Heelan suggests that the cultural worlds of antiquity and the Middle Ages presupposed space which was full, finite, and curved, and that this was their visual experience. "The establishment of cultural space as infinite and Euclidean was the achievement of the European Renaissance. Hyperbolic [curved] visual space was abandoned as the source of norms for realistic description, and in its place, Euclidean [straight] norms, connected with the scientific image of classical science, became entrenched." He proposes that Euclidean and hyperbolic visual space are complementary, that we have generally lost the latter and that to choose to recognize it would tremendously broaden our sense of our place in the world: "The inclusion of hyperbolic visual space within an enlarged realism would make a positive statement about the primacy of perception and the centrality to perception of physical embodiment in the [material fabric of the] universe. It would state that persons as perceivers are genuinely part of reality, like wave lengths, entropy or [possibly] black holes.

"... Such an enlargement of the notion of human realistic experience would also help us to recover the cultural 'myths' our ancestors lived by ... and to understand, relate to and even to construct 'natural' environments to which we could respond, like our Greek and medieval ancestors, without the real or assumed need of a particular technological shell."

Heelan's study shows that there is no perception that is not structured by conceptual and physical intermediaries that shape the way we perceive, and that such structures can be the artifacts of human invention as well as the biological basis of our embodiment: our own bodies and the evolutionary history of our race.

10 Particles?

Let us now look more closely at the "contents" of space-time, those little, elementary, impenetrable particles, and see just how elementary and impenetrable they may be.[1] In 1859 Plucker discovered that the cathode of a discharge tube he was working with was emitting invisible rays, detectable only as a green fluorescence in the side of the tube. He called them cathode rays. They are what causes the glow on the screen of a TV tube. In 1897, after a brilliant series of experiments, J. J. Thomson announced that these rays consisted of beams of tiny particles, later called electrons, each of which carried a fixed negative charge and was almost two thousand times lighter than the hydrogen atom, the lightest atom. Apparently, these atoms were *not* the smallest pieces of matter. In the previous year, Henri Becquerel had shown that uranium salts send out rays far more penetrating than X-rays. These rays were called alpha rays. This was the discovery of radioactivity. In 1903 it was found that when a piece of radon was placed in a sealed and evacuated glass container and left for a while, helium gas collected in the container. The alpha particles were in this way shown to be helium atoms ejected from the radon atoms. Apparently, then, atoms were *not* immutable.

Thomson proposed that an atom might consist of a heavy positive charge, equal to its atomic weight, with an equal number of the tiny, newly discovered, negatively charged electrons embedded in it like plums in a pudding. A rival theory was that the atom may be like a solar system, with all the mass and positive charge concentrated in a nucleus, like the sun, and the electrons flying around it like planets, kept in orbit by the attraction between their negative charge and the positive nucleus. In 1911, Ernest Rutherford realized that he could distinguish between these two models experimentally using Becquerel's alpha rays. He fired these alpha particles at a thin gold leaf. If Thomson's model were correct, the massive positive part of the gold atoms would fill the entire gold leaf, and the alpha particles would not be able to penetrate it. If the solar system model were correct, the gold nuclei would be arranged in the gold leaf like rows of billiard balls with huge gaps in between. In this case, if a beam of alpha particles was fired at the gold leaf, most of them would pass straight through and a few of them would be scattered in all directions as if they had hit a very heavy object. Rutherford found that the alpha rays did indeed behave according to the solar model. These scattering experiments formed the prototype of all future experiments in elementary particle physics: throw a beam of particles at a target really hard and see what bits you can break off, or how your original particles are scattered by the target.

Thus the atom now began to look very different. Instead of being an impenetrable lump, it had a minute impenetrable nucleus carrying almost all its mass and positive charge. Around this nucleus a number of electrons equal to its atomic number were orbiting. The diameter of the nucleus was found to be one ten-thousandth of the diameter of the whole atom. If you were to imagine the nucleus as being the size of a marble in the center of the Houston Astrodome, then the electron would be like a flea flying around the roof of the dome. In between the electron and the nucleus is "empty" space. The nucleus of the hydrogen atom is just a single positive particle, the proton. The nuclei of heavy atoms contain a number of protons equal to their atomic number.

In 1932, a third particle, the neutron, very similar to the proton but with no charge, was discovered. At this point, a rather pretty explanation of the atomic table begins to emerge. The nucleus consists of a number of positive protons equal to its atomic number together with a number of neutrons to make up almost the remaining mass of the atom. The nucleus is surrounded by a cloud of electrons whose negative charge balances the positive charge of the protons. In the decades to follow, the principles of chemistry were to a very large extent explained by the interaction of the clouds of electrons of various atoms combining to form molecules. The phenomena of color, of smell, and of hardness, as well as the variety of substances and their combinations, all stem from these electron clouds surrounding almost "empty" space: "empty" except for the tiny nuclei made up of protons and neutrons.

There remained the question of how a charged particle could in fact stay in stable orbit, because according to classical theory a moving charge should constantly radiate electromagnetic energy (light). Thus, as it lost energy, it would quickly spiral into the center. Bohr's solution to this problem was one of the major stepping-stones to quantum mechanics, which we will look at later in the chapter. Bohr took the audacious step of simply assuming that, for some unknown reason, electrons were forced to orbit the nucleus in discrete, prescribed orbits, without radiating any light. He then proposed that occasionally an electron might jump from one orbit to another, radiating a specific amount of energy (a quantum of energy) as it did so. Bohr calculated the spectrum of light we would expect to see emitted from a hydrogen atom if his model were true. His model was brilliantly confirmed by the agreement of his calculations with the observed spectrum.

Now let us return to the nucleus. The size of a proton and neutron is approximately only one-tenth that of a medium-sized nucleus, so they are quite tightly packed in there. We might say, ah! there are our hard, impenetrable particles, the protons and neutrons, tightly packed to form nuclei. Unfortunately, the story does not end there. The first problem is that if the protons are all positively charged and confined to a very small volume, why does the repulsive electric force of the positive charges not force them apart? There must be an even stronger force holding them together—this is known as the strong nuclear force. In 1935, Hideki Yukawa proposed that there should be an additional particle, later called the pi-meson (pion), responsible for carrying this strong nuclear force. Exchange of the pi-meson between two protons or a proton and a neutron would bind them together rather as two people tend to stay together if they are throwing a ball backward and forward between them. The pi-mesons would then form a cloud in the nucleus around the protons and neutrons, binding them together—a nuclear glue. In 1946, particles fitting the required characteristics were found in cosmic rays. Thus there were now four basic types of particles making up all atoms, the proton, neutron, pi-meson, and electron. The picture was still quite simple.

However, in the 1950s, physicists began to build huge machines, the cyclotrons, to accelerate particles to tremendously high energies. These particles could then be used in scattering experiments, to investigate the forces between them, like Rutherford's scattering experiments on gold leaf which lead to discovery of the nucleus. When these machines began to be used, new, heavy, and extremely short-lived particles were found. By the early 1960s there were literally hundreds of such particles. Some of these lived long enough to be detected by the tracks they left in bubble chambers; others were so short-lived that they could be detected only indirectly from the tracks left by the particles they decayed into, or just by bumps in the scattering data.

By the mid-sixties physics was in rather a mess—there was a plethora of particles, sometimes lovingly called the Particle Zoo, and no apparent way of understanding their relationship or why there were so many of them. Gradually, however, physicists began to detect some patterns in this chaos. (They were even able to predict the existence of a new particle as yet unknown which was subsequently detected—the omega minus. The omega minus was discovered when physicists recognized that there was one missing in what would otherwise be a nicely symmetrical arrangement of eight particles; on looking for a particle with just the properties of this missing link, they found it.) Physicists looking for elementary particles believe that where there are patterns there is underlying structure, and where there is underlying structure there is a more elementary level of building blocks. The really "elementary" particles cannot have structure. Thus the quarks were born; particles more elementary than even the proton and neutron—three quarks made one proton. For a while, simplicity seemed to have returned. However, it appears that in order for quarks to have the energies they need for them to bind together into a proton, their size needs to be one ten-thousandth the size of a medium-sized nucleus. Thus, the nucleus now appears to be made up of mostly "empty" space with minute quarks flying around in it. To use our marble analogy again: if the quarks are the size of a marble they will be flying around in a space anywhere from the size of a large barn to the size of the Astrodome. There is not even a heavy center to the nucleus, as the nucleus was for the atom. The cloud of orbiting electronic charge defines the periphery of the atom. But the periphery of the nucleus is defined by nothing; it is simply as far as a quark can go. A nucleus is simply a buzz of flying quarks, just as a cloud of gnats is simply gnats flying within a certain perimeter. But even the smallest nucleus, a proton, is like three gnats, or rather bumblebees, flying in open space the size of a large barn. Thus the nucleus becomes mostly "empty" space within the mostly "empty" space of the atom.

Is there any end to this? Some physicists working with quarks think they have reached "rock bottom" but give no very strong reason for their belief. As Heinz Pagels says, "No physicist I know would be willing to bet much on it." There are now five known quarks, rather than the original three, and almost certainly one more. Thus there are more quarks now than there were heavy particles in 1950. And whereas the protons and neutrons were bound together by one type of particle—the pi-meson—quarks theoretically are bound together by eight different types of particles, the gluons. Many physicists think there are already too many quarks for them to be truly elementary and are already talking about prequarks. Some physicists think there will be no limit down to the level of bubbles in space-time.

Space is Not "Empty"

But now let us look back at our description of the atom: a negative electron charge cloud surrounding "empty" space in which is a tiny nucleus which consists of tiny quarks flying around in "empty" space. Now, what about this "empty" space—is it really *empty*, void? The answer, according to physicists, is not at all. To understand this we must go back to 1928 and another extraordinary discovery. Paul Dirac was trying to form mathematical equations describing the motion of an electron in an electromagnetic field. He wanted these equations to bring together the newly developed quantum theory with Einstein's relativity theory. But he found in doing so that the equations had a strange characteristic: for every solution depicting the motion of an electron with positive energy, there was also a solution with negative energy. Since a particle cannot have negative energy, Dirac proposed that what these solutions really represented was a vast ocean of "virtual" electrons. He proposed that it might be possible for one of these "virtual" electrons, if it were supplied with enough energy, to jump up into a positive-energy or real state. This would leave behind a hole in the negative-energy ocean. Dirac proposed that such a hole would appear to us as a *positively* charged particle having positive energy and otherwise exactly like the negatively charged electron. Such a positive electron—called the positron—was very soon found.

The discovery of a particle predicted by theory is a very rare event—we have mentioned Yukawa's prediction of the meson in 1935 and the prediction of the omega minus in 1962 and their subsequent discoveries. The only other such case is the prediction in the late 1970s of the gluon—the particle that carries the force between quarks (properly termed the intermediate rector boson) — and its tentative discovery in the early eighties. Thus, such confirmations of a theoretical prediction are taken to be tremendous confirmations of the underlying theory. In the case of the discovery of the positron, physicists were already convinced of the correctness of Dirac's theory because it explained some fine details in the motion of the electron in the hydrogen atom which were not yet explained. So the discovery of the positron gave very strong credence to Dirac's proposal of an ocean of virtual particles existing all the time in the so-called vacuum and detectable only when one of them was knocked up into a "real" state leaving behind a hole. The particle and its hole would be detected as an electron-positron pair. This was the discovery of antimatter, the positron being the antiparticle to the electron. Later, all types of particles came to have their corresponding antiparticle, and therefore their corresponding ocean of virtual particles.

Dirac's work in bringing together quantum theory and relativity was extended over the next twenty years, and another way of understanding these virtual particles was conceived. In this view, the whole of space was filled with fields like the electromagnetic field, with one field for each type of particle. The degree of vibration of one of the fields at any point in space told how many real particles were there, if we wish to use the particle language. Thus, we have now two equivalent ways of thinking about the vacuum. We can think of it as oceans of invisible particles, of which the particle-antiparticle pairs we detect are like waves on the surface. Or we can think of it as filled with fields—infinitely many in the smallest region of space—all vibrating slightly, the vibrations corresponding to the virtual particles.

There is debate amongst physicists as to which of these two pictures represents "reality." Is "reality" particles whose probability of being in a particular place is determined by a mathematical, but not real, quantity which is the field? In this case, the vacuum is filled with infinitely many virtual particles which we cannot see but which can become "real" if they receive a sufficient jolt of energy. Or is "reality" infinitely many actual fields spread throughout all of space, whose low-level vibrations fill the vacuum and whose higher vibrations determine the presence of what we call particles? Which you choose seems to depend on whether you are working in the area of particle theory or field theory. But we are touching here on problems of interpretation at the very foundation of quantum mechanics, which we will deal with later.

The point for now is that, whichever picture you choose, space is no longer empty. It is so full that according to some calculations one cubic centimeter (a thimbleful) of space contains, in its ocean of virtual particles, as much energy as the energy of all the "real" mass in the universe. [2] How then shall we now consider our atom—our little elementary impenetrable ball? Perhaps we could say that its surface, determined by the electron orbits, appears in already full space. Then the surface of the nucleus also appears in already full space, and the same for the surface of the quark. . . . down to what? Surfaces, bubbles, of space-time itself? Bubbles, within bubbles, within bubbles . . . formations, within formations, within formations in the fullness of space?

The Particle-Wave Duality

Let us now look into how it was that physicists began to associate fields or waves with particles. We shall see that this introduces a profound ambiguity into the concept of matter and raises questions as to whether such a concept can in any sense apply to the reality of the minute world. In fact, it raises profound questions as to how even the concept "reality" can be applied at this level.

First I must explain a few simple ideas. A particle is a lump of matter (or energy, matter and energy being equivalent). The characteristic of a lump of matter which leads us to call it a particle is that it is all concentrated in a small region of space at a particular time, or we could say it is localized in space-time. In contrast, a wave or field is also energy, but it is spread out throughout space-time. For example, if two people hold one end of a rope and one of them shakes it, a form moves down the rope. The wave is the whole motion of the rope as it changes in time along its length. Or if a gust of wind catches the edge of a field of wheat, sending a ripple across the field, then the "wave" is the motion of the entire field. Clearly, then, particle and wave represent quite opposite ways of distributing energy in space and time. We could ask about the position of a particle, but it would be meaningless to ask about the position of a wave. On the other hand, we can ask about the frequency of a wave (the number of times it vibrates per second), but it would be meaningless to ask about the frequency of a particle, a lump of stuff.

The main characteristic by which we can distinguish wavelike behavior from particlelike behavior is the phenomenon of interference. If you drop two stones in a pond, waves spread out in circles from each stone, and where the waves cross you see patterns of interference. Where two troughs meet you see a trough twice as deep; where two peaks meet you see a peak twice as high; and where a peak and a trough meet they cancel out and the water remains flat. The reader can observe an interference phenomenon very simply: squeeze your fingers together so that there is a paper-thin crack between them. Put your eyes close to this crack and look at the sky or a piece of white paper through it. You will see many light and dark bands.

In 1802 Thomas Young performed a simple interference experiment which clearly showed that light is wave motion. He allowed sunlight to pass through a single pinhole onto a screen with two pinholes in it, beyond which there was another screen. Instead of seeing two overlapping circles of light, one from each pinhole, he saw many bands of light and dark, clearly indicating that interference was taking place and that therefore light must be considered wave motion. Two beams of particles could not produce an interference effect; in particular there could be no dark bands, for particles could not cancel out.

By the turn of the century, physicists were firmly convinced that light was, in fact, wave motion of the electromagnetic field. There were just two puzzling phenomena that could not be explained on this basis. The first was the way light energy radiated as the temperature of a "perfectly black" object was gradually raised (a phenomenon known as black-body radiation). In 1900 Max Planck proposed that the only way to explain the characteristics of such radiation was to assume that for some reason light was being emitted from the body not continuously but in minute chunks or quanta. Planck himself did not consider these quanta or light particles to be "real." He thought he had discovered a mathematical trick. Light had, after all, only very recently been proven to be electromagnetic wave motion. In fact, when, several years later, Planck was sponsoring Einstein's appointment to the Prussian Academy of Sciences for his work in relativity, he asked that the officers of the Academy overlook Einstein's "youthful folly" in proposing that light quanta are real. Once again we find "I'll see it when I believe it." However, in 1905 Einstein provided a brilliant explanation of the second unexplained puzzle—the photoelectric effect—in which he did indeed propose that light actually was behaving like particles. In the photoelectric effect, when light shines on a piece of metal an electric current of electrons is emitted. The relationship of the magnitude of the electric current emitted to the intensity and color of the light shining on the metal could not be explained by assuming classically that the metal gradually absorbed light radiation until it had absorbed enough to kick off an electron. But it *could* be explained very precisely by assuming that the metal absorbed light in chunks, quanta, or particles, all at once. Thus, to everyone's astonishment, light was now, in some circumstances, behaving as if it were a stream of particles, while in other situations it clearly behaved like a wave motion.

Now let us return to matter. This is made up, so we believed, of little particles known as atoms, which are themselves made of little particles known as protons, neutrons, and electrons. However, following Bohr's extraordinary success in explaining the emitted light spectrum of the hydrogen atom, there remained the puzzle of *why* the electrons were confined to certain orbits around the nucleus. In 1923 Louis de Broglie made the suggestion that just as light waves could sometimes behave as particles, so perhaps electrons could have a wave nature as well. The restriction of orbits in the hydrogen atom would then be explained by saying that only those orbits were allowed which could contain an integral number of half wavelengths of these electron waves. This can be illustrated by the fact that if you tie one end of a rope to a tree and shake the free end, the rope will vibrate with an exact integral number of half wavelengths (the distance between two peaks) between your end and the tree—these are the harmonics of the vibrating rope. There may be a half or one wavelength, or one and a half, or two, or three, and so on, but there cannot be one and a third wavelengths. Similarly, when you pluck a violin string, only certain frequencies can be heard—the harmonics—which correspond to the length of the string's being one-half wavelength or integral multiples of this. De Broglie's suggestion was that the allowed orbits of the electron in the hydrogen atom were proportional to the "harmonics" of its matter waves. The "wavelength" of an electron matter wave could then be calculated. In 1923 Davisson and Germer fired a beam of electrons through a thin layer of gold (which acted like a multiple slit) onto a photographic plate. The plate showed clear interference patterns. The beam of electrons was behaving like waves. The wavelength of the "matter waves," which could be calculated from the width of the interference bands, was in agreement with de Broglie's calculation for the wavelength of the electron's "matter waves."

At first one might think that these results showed that a beam of electrons can actually spread out into "matter waves." Consequently, Erwin Schrödinger developed an equation describing the changes over time of these matter waves, which looked very similar to the form of Maxwell's classical equations for the radiation of electromagnetic (light) waves. And the solutions to Schrödinger's wave equation for motion of an electron in a hydrogen atom corresponded very precisely to the atomic orbits of Bohr. However, when we try to visualize what kind of "matter" these matter waves could be waving, we have serious problems, especially when we consider the wave associated with a single electron.

An interpretation of the waves was provided by Max Born, who suggested that we should suppose that electrons remain particlelike but that they have associated with them a wave whose amplitude at a particular point determines the *probability* of finding the electron at that point. Born's view was, then, that the electron always remained a particle and that the "waves" were simply mathematical devices physicists used to find the probability of finding a particle in a particular place. However, as we shall now see, there are problems with this interpretation as well. If we try to think in terms of everyday images of "waves" and "particles," then to say that the electrons become smeared out and behave like a wave in the interference experiment, or to say that the electrons continue to retain a characteristic like a classical particle, both seem improper. We simply cannot think about electrons adequately in either of these classical pictures.

In order to see this more clearly, I will describe a well-known "thought experiment," the double-slit experiment.[3] Suppose we imagine three two-part experiments: In the first we have an armor-plated screen with two slits in it. We fire a machine gun at the screen, and on the other side of the screen is a target. If we open only one of the slits, the distribution of the bullets will look like Figure 3A. If we open both slits the distribution of the bullets will be a simple summation of the distribution from each hole separately, as in Figure 3B. Now suppose we do a similar experiment in which, instead of a machine gun firing bullets, we now have a barrier with two slits in a tray of water and a vibrator on one side of the barrier to form water waves. Suppose in place of the screen we have a row of corks to measure the amplitude of the waves. Again, if we open only one hole we will see a pattern on the other side of the hole as in Figure 3C. But now, if we open both slits, because of interference of the two waves, the pattern of the corks will look like Figure 3D. In the third experiment, we use a source of electrons instead of bullets or waves, and a metal plate with two slits in it with a photographic emulsion screen on the other side. The patterns we see on the emulsion with one slit open and with two slits open will be similar to the patterns we found with water waves and *not* with bullets. (Figures 3E and 3F).

Now, if we are using a stream of large numbers of electrons, we might say, well, somehow the individual electrons cooperate to produce a kind of wave motion in the overall stream itself. So let us make the intensity of the stream of electrons so low that only *one* electron is passing through a slit at any time. As each electron strikes the emulsion, we will see a small black dot appear—this would seem to indicate that, at that moment, the electron is actually at that point

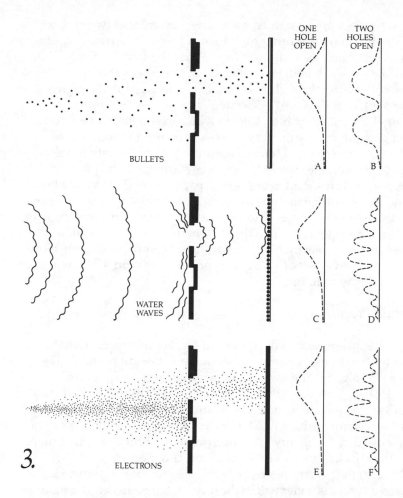

ONE HOLE OPEN

TWO HOLES OPEN

BULLETS

A

B

WATER WAVES

C

D

3.

ELECTRONS

E

F

where the black dot appears, i.e., it is behaving like a particle. But as more and more black dots appear on the screen, the pattern they form is the *interference* pattern; it is *not* the pattern we would get if the electrons were behaving like little bullets all the way from the source through the double slit to the screen. So if we have been imagining the electrons as little tiny lumps of stuff, like bullets leaving the source, passing through one of the slits and hitting the screen, moving all the while along a definite path, then we have been mistaken. If the electrons had been doing that, then the pattern that would have begun to appear on the emulsion would have been like the pattern in the bullet experiment, not the water wave experiment. Furthermore, it appears that each electron must somehow "know" that two slits are open rather than one, in order for it to make an appropriate contribution to the blackening emulsion. Thus, either the electron is somehow "spread out" enough to detect the fact that the other slit is open (the one it is not passing through, if we think of it as passing through a definite one), or the "wave" associated with the electron must be detecting the actuality of both slits, in which case it has some physical reality and is not merely the way physicists compute probabilities.

The reader might find himself asking, "Yes, but what then is *happening* between the slits and the screen? Are there real physical entities which correspond to our terms 'particles' and 'wave'?" Now we again are beginning to come up against the tendency of the human mind, discussed in Chapter 3, to assume that because we can speak of something, it is therefore a real thing. Whether or not we want to say that there is a *real* objective, physical thing corresponding to the terms "particle" or "wave" or both is a matter of interpretation which we will discuss in the next chapter.

Quantum Weirdness

The so-called weirdness of the quantum level of reality lies right here: that we wish to think of electrons as ordinary classical particles like tiny billiard balls, having a definite trajectory, and yet quantum mechanics keeps telling us that we cannot think of them like that at all. We can see this weirdness again from another point of view—Heisenberg's "uncertainty principle." This is a direct and inevitable result of the logic of quantum theory. The uncertainty principle says that there is an *inherent* indefiniteness in the position and momentum of a particle, such that the more precisely the position of the particle is narrowed down, the wider will be the indefiniteness of the momentum of that particle. This uncertainty shows up in an inability to measure both the position and the momentum as precisely as we wish, but it is not only a characteristic of the measurement process. It is a characteristic of the nature of particles at the quantum level—an inherent fuzziness of trajectory which makes it not possible *even in principle* to determine both where they are and whither they are going with complete accuracy. It is this principle that dictates that we simply should not try to visualize that after passing through a slit the electron *is actually* headed toward a particular point on the photographic screen.

Perhaps you might now be inclined to say, "Maybe there are deeper levels of reality from which other factors act on the electron, factors that we are not now aware of and cannot be aware of until we find an altogether different way of making measurements. Possibly these factors (hidden variables, as they are called) do determine a real, objective trajectory for real, objective particles." This was the view that Einstein took all through decades of back-and-forth discussion with Niels Bohr. There may well be deeper levels of reality than the quantum level, but are there hidden variables? One of the most dramatic results of the physics of recent years is the theorem of John S. Bell and its subsequent very clear experimental verification. This work essentially shows that there *cannot* be "hidden variables," thus ending the last attempt to hold onto our image of a real, objective electron traveling along a real, objective path. [4] As we will see in the next chapter, physicists are still trying to propose various possible interpretations of "what happens" between the double slit and the screen. However, whatever the electron is, if we can describe it at all, it is nothing like a little ball of stuff.

Before we go on to consider these interpretations, it is necessary to consider one more aspect of the weirdness, that is, the behavior of the probability waves when any interaction with a measuring instrument occurs. Let us return to the double-slit experiment. Before the black spot appears on the screen, the electron "is" heading toward the screen with various probabilities of hitting it in each point—the greater probabilities being where the bands are blacker. These probabilities are the wave aspect—the probability wave. Now, when the black dot appears—that is, when the electron hits the screen in a particular spot, remanifesting as a particle—then the probability wave abruptly and instantaneously changes to being 100 percent that the electron is where it is. If there is, indeed, any element of physical reality corresponding to the probability wave, then such sudden changes of form when a measurement is made is, certainly, strange.

Let us look just a little further into the possible relation between the "particle," its associated probability "wave," and the act of measurement, in a thought experiment proposed by Renninger.[5] Let us suppose that we have a source of electrons which fires once every ten seconds in a random direction. Suppose we place this source in the center of a hollow sphere covered with photographic emulsion, and that it takes an electron two seconds to travel from the center to the surface of the sphere. Every ten seconds a black spot appears randomly on the sphere, two seconds after an electron was fired from the center. Between the time an electron leaves the center (time: 0 sec) and the time it reaches the emulsion (time: 2 sec), according to quantum mechanics, the electron is not traveling in *any* specific direction. Therefore, the probability wave is an even distribution over a sphere which moves outward with the electron. However, at time: 2 sec, we know precisely where the electron is (at the black dot); therefore the probability wave was instantly reduced from the sphere to a dot. This reduction of the wave from an infinite number of possibilities to one actuality is similar to the case of the two-slit experiment where the wave was reduced from many possibilities, determined by the bright and dark interference bands, to just one.

We might say, "Well, it is the interaction of the particle with the measuring instrument (the emulsion) that causes this change in the form of the wave." But now consider a case where there is no physical interaction of the particle and the measuring instrument. Suppose we place a half sphere of emulsion exactly halfway between the center and our first sphere. It will take one second for the wave to reach this half sphere. Suppose there is no mark on this emulsion after one second; then we know that the electron went through the other half of the sphere in its journey to the outer sphere. The probability wave has now changed from being a full sphere moving along with the electron to being a half sphere. The probability of the electron's being in that half where the inner emulsion is has been reduced to zero merely because the observer did NOT see an interaction between the electron and the inner hemisphere at time: 1 sec. Here it appears to be the observation rather than any physical interaction that changes the probability wave.

To summarize then: an electron appears to manifest properties of a particle but also to have associated with it a probability wave. The form of this probability wave appears to change instantaneously when there is interaction with a measuring instrument, even if the "particle" does not itself interact physically with the instrument. Whether the electron shows more particlelike or wavelike characteristics depends on how we set up our experimental situation. If we are doing an experiment like Davisson and Germer's, looking for wavelike properties of the electron, then we will find them. If we are doing experiments to find the position of the electron, a particlelike property, then we will find it. But an experiment cannot be set up to ask these two questions simultaneously. This is the origin of Bohr's complementarity principle, to which we will return in the next chapter.

This is unquestionably the most difficult point for anyone, layman or professional scientist, to grasp in all of science today. Our mind continually flips back and forth between two concepts or images familiar to us from our large-scale world: the image of a particle, a concentrated little lump, and a wave, a vibration or pattern which spreads out through an extensive space. We think, "Of course it is a real particle, we just don't *know* where it is, the wave gives the probability of finding it." For a moment we feel satisfied, and then we remember the double-slit experiment with single electrons: they simply *do not* behave like bullets. It is as a wave that they seem to have some kind of physical interaction with the double slit. Then we remember the experiment with the half sphere in which it seems to be merely our knowledge that changes the wave function. So we return to thinking that the wave is simply a mathematical device. Our mind chases around in this circle until we realize that our images of familiar objects or patterns of everyday life simply cannot be applied to the electron. We have no concepts available to us from everyday life to picture to ourselves what is "really going on."

You may even be beginning to think by now that you really don't understand what is going on at the quantum level of reality at all. If this is the case, then be heartened by knowing that you are at least beginning to understand the difficulties that physicists have faced for more than half a century. As Richard Feynman has said, "Do not keep saying to yourself if you can possibly avoid it, 'But how can it be like that?' because you will go 'down the drain' into a blind alley from which nobody has yet escaped. Nobody knows how it can be like that."[6] The response to the tremendous open-endedness of reality, to its refusal to be bounded by classical description, is often simply to ignore it, even on the part of traditional physicists. As Davies says, "These statements are so stunning that most scientists lead a sort of double life, accepting them in the laboratory, but rejecting them without thought in daily life." Yet there is a very profound effect on our daily lives when we realize that physics has finally penetrated to the very core of material, objective reality and found nothing that can ultimately be said to exist beyond our measurement of it. Objective reality—that which is real independently of its being known, that which is behind appearances and is more real than appearances—this objective reality has slipped beyond concept, even the concept of existence or nonexistence. Those impenetrable little lumps of stuff—the elementary particles—held to be the ultimate building blocks of everything have certainly vanished. And in their place we have "appearances" which depend on what kind of measurement we are making, what kind of observation or perception is taking place. Whether there is anything behind these appearances, and what it might be, are matters of interpretation which we will discuss in the next chapter.

11 Interpretations: The Possible and the Actual

In the previous chapter we saw that quantum mechanics requires the introduction of a wave or field associated with each particle. We saw that mathematically this wave gives the probabilities of finding various possible results when we make a measurement (of position, momentum, spin, etc.) on the particle. We also saw that the field changes abruptly when that measurement is made—the probabilities of various results are reduced to the observed actuality of one particular result. The probability wave seems to have very real consequences (the interference bands), and it is an essential part of the mathematics of quantum mechanics. Yet it seems to change instantaneously when we make an observation. This violates principles of classical mechanics and relativity as well as our intuition. Now, the question we might naturally ask, which was indeed asked by many physicists at the time quantum mechanics was being formulated, is: is there an element of physical reality which this wave function describes, or is it *merely* a computational device? More particularly, what precisely "happens" in the "real world" when the probability wave is reduced?[1]

The Copenhagen Interpretation

Many physicists do not like to ask these questions. Quantum mechanics works beautifully beyond our imagination. Because of quantum mechanics we have lasers, superconductors, atomic energy, quartz watches, so why worry about the foundations? Let us simply say that we do not know.

This is the popular version of the Copenhagen interpretation, which was taken for granted from the late 1920s to the late 1960s by all but a very few physicists. The Copenhagen interpretation says, Until we make a definite observation of the world, it is meaningless to ascribe to it a definite reality. It was formulated by the great elder of quantum physics, Niels Bohr, and extended by another leading quantum physicist, Werner Heisenberg. Bohr's approach was to say: we live in a relative world that behaves classically; our observations at the level we make them occur in large machines obeying classical mechanics. Therefore, what happens at the quantum level is simply not knowable by us, and we should be prepared not to try to make any statements about it. He did not say there was no definite reality, simply that we could never know it completely at the quantum level. All we can know is what happens at our relatively large scale.

However the Copenhagen interpretation has been taken by a generation of physicists to mean "we cannot know anything beyond the pragmatics, so there is no point in thinking about it." This was by no means Bohr's intention, or the way he himself responded to his interpretation. He thought about it very deeply throughout his life.[2]

Bohr formulated the principle of complementarity: that it is in the very nature of reality that there exist qualities or aspects that are impossible for human knowledge to grasp simultaneously. For example, the wave and particle natures of the electron are two such aspects. Bohr extended this principle far beyond the problems in measurement of the electron to every realm of life, biology, psychology, and philosophy.

According to Max Jammer, who has made a very detailed study of the origin and development of the concepts of quantum mechanics, Bohr was strongly influenced by the Danish philosopher Hoffding and also by the American psychologist William James, who in turn was strongly influenced by the writings of Renouvier. According to Jammer, "Renouvier was one of the earliest modern thinkers who questioned the strict validity of the causality principle as a regulative determinant of physical processes. Renouvier proposed a phenomenalism according to which all that we immediately know is but a particular phenomenon or 'representation.' Every representation has a two-fold character, it is a 'representing' and a 'represented'; it is an experience of something and it is something *experienced*. Realism is erroneous, Renouvier contends, insofar as it assumes that the object can be divorced from its representation and idealism is erroneous insofar as it assumes that there exist representations with nothing to represent." And in turn Hoffding recognized the arbitrariness of the division between subject and object: "Man cannot without falsification conceive of himself as an impartial spectator or impersonal observer; he always necessarily remains a participant. Thus man's delimitation between the objective and the subjective is always an arbitrary act and man's life a series of decisions. Science is a determinate activity and truth a human product, not only because it is man who created knowledge but because the very object of knowledge is far from being a thing ready-made from all eternity."[3]

William James was particularly influenced by Renouvier, whose philosophy brought James out of a depression in 1870 and to whom James dedicated his last work, *Some Problems of Philosophy*. And Bohr, in turn, was influenced by James. In an extraordinary interview recorded the day before he died, Niels Bohr recalled the influence on the development of his interpretation of quantum mechanics of a passage in a chapter by James called "Stream of Thought."[4] In this chapter James insists that "thought and thinker, subject and object are tightly coupled. The objectivization of thought is impossible . . . Our mental reaction to every given thing is really a result of our experience in the whole world up to that date." And the key passage which seems to foreshadow Bohr's theory of the behavior of electrons in an atom as well as the principle of complementarity: "Like a bird's life [thought] seems to be made of an alternation of flight and perchings. The rhythm of language expresses this, where every thought is expressed by a sentence and every sentence closed by a period . . . Let us call the resting places the 'substantive' parts and the places of flight the 'transitive' parts of the stream of thought." Bohr's comment on this chapter in that last interview is: "In a clear manner it shows that it is quite impossible to analyze things in terms of—I don't know what to call it, not atoms, I mean simply, if you have some things . . . they are so connected that if you try to separate them from each other, it just has nothing to do with the actual situation."[5]

I have quoted these passages at length because I feel it is important to understand the context in which Bohr himself seems to have meant his interpretation. Like Einstein, Bohr was a man of tremendous breadth, humor, a large heart, and a love of life. He loved to ride his bicycle and play soccer; in his youth he played for the Danish National second team. During the war he insisted on staying in Denmark, which was an occupied country, in part so that he could supervise the use of his physics institute as a route for Jewish scientists escaping from Nazi Germany. He never divorced his principles of physics from his principles of life. He intended his interpretations of quantum mechanics to apply to life in a much broader way. As he himself said, "It is significant that . . . in other fields of knowledge, we are confronted with situations reminding us of the situation in quantum physics. Thus, the integrity of living organisms, and the characteristics of conscious individuals and human cultures present features of wholeness, the account of which implies a typical complementarity mode of description. We are not dealing with more or less vague analogies, but with clear examples of logical relation which, in different contexts, are met with in wider fields."[6] Jerome Bruner recounts a conversation with Bohr which showed the extent to which he felt the principle of complementarity governed his own life: talk turned entirely on the complementarity between affect and thought, and between perception and reflection. [Bohr] told me that he had become aware of the psychological depths of the principle of complementarity when one of his children had done something inexcusable for which he found himself incapable of appropriate punishment, 'You cannot know somebody at the same time in the light of love and in the light of justice.'"[7] The ability to live and delight in the tension of holding apparently opposing views without having to grasp either of them, manifest in the lives of both Bohr and Einstein, is sometimes said to be the mark of wisdom.

Bohr's view of the probability wave was, then, that it provides a means of calculating the probability of finding certain results when we make a measurement. For example, in the double-slit experiment, the probability wave can tell us the probability of finding an electron in a particular area of the photographic plate after it has passed through the double slit. It *cannot* tell us what the electron is doing or even *is* between the slit and the screen. Between the slit and the screen, reality simply does not have the definiteness that we think it should have when we think about it in our usual way. Therefore, we must change our modes of thinking about reality at this level, not demand that it conform to our usual ways of thinking based on a world of objects. The principle of complementarity was Bohr's way of thinking afresh. For Bohr it is simply meaningless to expect any more definite answer to the question, "What is behind the appearance?"

Heisenberg, Bohr's colleague in the early development of quantum mechanics, at first agreed with Bohr's view. However, in later years, he did speculate on how to speak of what is behind appearances. He proposed that although we cannot strictly speak of what "happens" between two observations, we can say that the system, the "object" of observation, must be in a state of possibilities or tendencies. He sometimes also called these possibilities "potentia," thus aligning himself with Aristotle's view of how matter takes on specific forms. Heisenberg suggested that the abrupt change from "potentia" to actual takes place when the "object" interacts with the physical measuring instrument and that this happens independently of the observer's knowledge. At the same time, Heisenberg agreed with Bohr that the division of the whole setup into "object" and "measuring instrument" was arbitrary, depending on how the observer chose to make this division. Thus the whole system, object plus measuring instrument, is in a sense self-reducing its potential to actuality.[8]

Karl Popper, independently, proposed an almost identical view, calling the probabilities (tendencies or "potentia" of Heisenberg) "propensities." For Popper these "propensities" were objective and physical.[9] For neither Popper nor Heisenberg did the change from potential to actual depend on the separate consciousness of an observer. However, for Heisenberg, with the physical change from potential to actual there is a parallel change in the observer's knowledge, and *this* change is reflected in the reduction of the probability wave. Heisenberg then rather curiously separates the change in the objective world from the change in the subjective world and says that the two happen together, that one is not the cause of the other: "The word 'happens' . . . applies to the physical, not the psychical act of observation and we may say that the transition from the 'possible' to the 'actual' takes place as soon as the interaction of the object with the measuring device, and thereby with the rest of the world, has come into play; it is not connected with the act of registration of the result by the mind of the observer. The discontinuous change in the probability function, however, takes place with the act of registration, because it is the discontinuous change of our knowledge in the instant of registration that has its image in the discontinuous change of the probability function."[10]

"Consciousness" Seems To Enter

However, if we continue to press the question, "But what precisely is it that reduces the potential to actual?" then we seem to need a way to consider the conscious act of registering the observation by the observer. The only alternative would seem to be to maintain a strict separation between materiality and mentality and then try to understand how they might interact. This is the approach of Karl Popper and John Eccles.[11] However, if we do maintain this separation we have to invoke some entirely new principles for the appearance of consciousness. As Eccles says, "Each self is a divine creation."[12]

John Von Neuman and Eugene Wigner argued that the end of any chain of events leading to a measurement is a human observer. Now, we can regard the physical body of the observer as a part of the physical measuring instrument. When we ask what in the whole act of observation is the crucial event, we are in the end led to the act of registration of the result of the measurement by the consciousness of the observer. Some physicists, therefore, with Wigner,[13] would like to say, "Yes, it is the act of consciousness of the observer which reduces the many possibilities to one actuality." This attitude is perhaps satisfying in that it begins to break down the duality between mind and matter. Some people even think it can explain psychokinesis.[14] It is a little discomforting in that it brings us rather close to solipsism: the world is not real, only a bunch of possibilities, unless observed by a consciousness. But whose consciousness? To understand how this question now arises, suppose there are two human observers, Mary and Tom. At 3:00 p.m. Mary tells Tom the result of an experiment that she observed half and hour ago. How should Tom regard the state of the system at 2:30 p.m.? He did not know then that Mary made the observation; so as far as he was concerned, the state of the system at 2:30 was not yet reduced from potential to actual. After Mary tells him of her observation, does this now change reality for Tom half an hour in his past? Or should we say that potential is reduced to reality for Tom not by Mary's actual observation, but by the possibility that anyone (for example, Tom) COULD have made such an observation during that half hour? In that case, since there is always the possibility that someone could make an observation, potential is always reduced to actuality, which would bring us to the "many worlds" view.

The most fundamental difficulty with this approach seems to be that we have no idea what we mean by consciousness in such a case. Normally we are conscious of something in the everyday world of bodies and tables. "I am conscious that my body has been refreshed by a sip of tea." Or we are conscious of the world of thoughts and dreams. If this consciousness has any separate reality apart from chairs and tables and dreams and measurements, then could it not also be conscious of the realm of possibilities before they become reduced to actuality and also be conscious of the act of reducing to reality? But if we are referring to a consciousness which is somehow tied to our mundane world, then how can this consciousness be said to affect the realm of potentials? We are in a double bind: either this "consciousness" reaches into the world of potentials, in which case why is it conscious only of that one which becomes actual? or it does not reach into this realm, in which case how can consciousness affect this realm so as to make one of the potentials actual?

Many Worlds

Because of this kind of tangle, we may think that consciousness obeys laws quite outside those of physics. This was the view that Wigner eventually came to. Alternatively, perhaps we had better drop the idea of consciousness altogether. This is what Hugh Everett and Bryce deWitt tried to do in their "many worlds" interpretation.[15] They suggested that instead of regarding the various probabilities as simply potential, we should regard them all as actual, each in a different world. All possible universes actually exist. Any time two particles interact, with several possible outcomes, each of those possibilities actually happen in some world. Since billions of particles are interacting at each moment, the universe is branching into further billions of actual universes every moment. These universes continue parallel to each other. Some differ by very little from each other. Some, those which branched off a long time ago, are so different as to be unrecognizable. The observer also branches each moment. He is, after all, part of the universe. But he does not know he has branched and continues to experience himself as one observer in one universe. How can this be? Consider two universes which differ by very little. They branched one-thousandth of a second ago and differ now only in the outcomes of a few particle interactions. Would the observer in one of these two universes not feel any sense of duplication, any shadow impressing itself on him? Or are these many universes, from the moment they branch, utterly separate? And again, who is this observer who experiences himself as one, when there are in "reality" almost infinitely many of him? We should note that in the many-worlds model the observer seems not to be important. That is to say, all these universes exist whether there is an observer in them or not. No observer is needed to reduce potential to actuality because all possibilities are already actual.

In view of what we know about conditions required for life and the way the universe may have started and evolved, a universe which would support life and consciousness would seem to be an extraordinarily improbable accident. The probability of there being consciousness has been estimated to be astronomically small. Is this just extraordinary good luck (or bad luck)? The many-worlds model would say something different: "All the other worlds are real, each with an equal status of existence. If life is very delicate, then most of these worlds are even now devoid of observers. Only ours and those very similar thereto will have spectators. In this case we have, by our very presence, selected the type of world we inhabit from among an infinite variety of possibilities."[16] This is known as the anthropic principle. To put this another way: it is as true to say that this particular universe is as it is *because* we are in it as to say that we are in this particular universe *because* it is as it is. All universes are equally real, yet you experience only this particular one. And what is the reason for the apparent special reality of this particular universe? That you are in it.

The Wigner and Everett models start at seemingly opposite ends of the mind-matter dichotomy. Wigner brings in consciousness as the key which brings down reality from the realm of potential. Then we lose track of consciousness when we try to pinpoint it. Everett starts with the reality of all worlds, independent of consciousness. Then we need to bring consciousness back in to explain how the many are lived in as this particular one. Both of them return to an insoluble dualism, because that is the premise they started from.

Wholeness and the Implicate Order

Now we turn to another model of the quantum level, in which consciousness and matter enter on a more equal footing right at the beginning—the model of David Bohm.[17] Bohm begins by recognizing all of what is as an undivided wholeness. Within this, man, in his thinking and language, makes distinctions; he divides things one from another. If man confuses his thought *about* what is with what is, then he and his world become fragmented.

However, the relation between thought and the reality that this thought is about is not one of correspondence of two separate substances or processes: "Clarity of perception and thought evidently requires that we be generally aware of how our experience is shaped by the insight (clear or confused) provided by the theories that are implicit or explicit in our general ways of thinking. . . .If we regard our theories as direct descriptions of reality as it is then we will treat differences and distinctions [in our theory] as divisions implying separate existence of [entities corresponding to] the various elementary terms appearing in the theory. We will thus be led to the illusion that the world is actually constituted of separate fragments." Bohm continues: "Relativity and quantum theory agree in that both imply the need to look on the world as an undivided whole in which all parts of the universe, including the observer and his instruments, merge and unite in one totality. A new form of insight implies that flow is in some sense prior to that of the 'things' that can be seen to form and dissolve in this flow. That is, there is a universal flux that cannot be defined explicitly but which can be known only implicitly, as indicated by the explicitly definable forms and shapes, some stable and some unstable, that can be abstracted from the universal flux. In this flow, mind and matter are not separate substances. Rather they are different aspects of one whole and unbroken movement."

Bohm then analyzes this unbroken movement as consisting of layer upon layer of more and more general levels of ordering which are hidden, enfolded, or implicate. Therefore, he terms this underlying flux the implicate order. Specific manifestation is the revealing or unfolding of parts of some layers of order out of the implicate by acts of immediate perception. This manifest appearance he then calls the explicate order: "In terms of the implicate order one may say that everything is enfolded into everything. This contrasts with the explicate order now dominant in physics in which things are unfolded in the sense that each thing lies only in its own particular region of space and time and outside the regions belonging to other things." To clarify the way in which the many explicates are enfolded into the implicate, Bohm uses a model based on the hologram. A hologram is a photographic plate which records the interference patterns of coherent laser beams, one of which is reflected off an object. The plate can in turn be read by a laser beam—producing a three-dimensional image of that object. The important difference between a hologram and an ordinary photographic image is that in an ordinary photo-graphic negative, each small area of the negative contains the information for an exactly corresponding small part of the object. However, in a hologram, each small area of the plate contains the information to reproduce an image of the entire object (the smaller the area, the fuzzier the image, but a complete image can be generated from even a very small section of the hologram). The hologram itself does not look anything like the object—whereas on a photographic negative you can literally see an image looking like the object.

The way in which the explicate orders are enfolded into the implicate order is, Bohm suggests, analogous to the way in which the holographic plate "enfolds" the three-dimensional image. Just as all of the image can be recovered from any small region of the holographic plate, so all of reality is enfolded into each small region of the implicate order. Bohm and neurophysiologist Karl Pribham have speculated that perception may involve a mechanism analogous to retrieval of a holographic image.[18]

According to Bohm, all of the fields of quantum mechanics—gravi-tational, electromagnetic, strong and weak nuclear force fields, fields of electrons, protons, light, and so on—all are enfolded into the implicate order which, because it is dynamic, he calls the holomovement. The extent of the holomovement is unknown and cannot be known: "Even the quantum laws may only be abstractions from still more general law, of which only some outlines are now vaguely to be seen. So the totality of movement of enfoldment and unfoldment may go immensely beyond what has revealed itself to our observation thus far." Clearly the holomovement could exist in more dimensions than the four of space and time. Bohm suggests that it does, on the basis of the apparent long-distance connectedness of the particles in the Bell experiments. Bell's theorem, and subsequent experiments corroborating it, appear to show that any two particles, once they have interacted with each other, will continue to remain correlated over long distances and times. Bohm draws an analogy with a fish in a fishbowl (the implicate order—a higher dimension) hidden behind a curtain. Two TV cameras are trained on the bowl and all we can see are the images from the camera on two TV screens (the explicate order in our four-dimensional space-time). When we see the two fishes apparently moving together, we are surprised. In fact, there is only one real fish seen from two different angles, and that is why their movements are correlated. In the higher dimension, there is only one particle, of which we see two aspects or projections in our lower dimension.

Finally, in similar fashion, Bohm suggests that, as well as matter, consciousness itself is enfolded with the implicate order, which is thus more primordial than either: "We are led to propose that the more comprehensive, deeper, and more inward actuality is neither mind nor body but rather a yet higher dimensional actuality, which is their common ground and which is of a nature beyond both." Mind and body emerge together as related projections from this higher-dimensional ground. Again we can use the analogy of the fishbowl. In this case, the two images we see on the TV screen are what in our four-dimensional world we think of as mind and matter. If we try to look directly at such higher dimensions in our normal state of mind, which is able to discriminate only large differences in space and time, we will be able to discriminate nothing. Therefore, since the appearance of things is due to discrimination, this higher dimension will appear to be no-thing. In fact, of course, we probably cannot directly experience the totality of the implicate order; we can simply discover or unfold larger and larger suborders of it into consciousness. Inanimate objects, living organisms, and conscious entities are all enfolded into the implicate order and become manifest by unfolding out of it. Should we say, then, that the implicate order is animate and conscious, or rather that it is inanimate and unconscious? Probably each of these descriptions can be applied only to a suborder; the implicate itself goes beyond such descriptions.

Bohm's theory brings together the fragmented parts of what is, providing the basis for a unified understanding of matter, life, and consciousness. Questions such as whether mind or matter came first and how one can affect the other do not need to arise here.

The implicate order model is less directly related to the detailed mathematics of quantum theory. It is therefore more difficult to talk about this model, in the terms that physicists prefer to talk, than to talk about Wigner's or Everett's model. It feels intuitively more grounded in our immediate experience, yet it still contains the strangeness of implying further dimensions beyond those which we directly perceive, namely, space and time.

Alex Comfort points to the analogy between superposition of quantum states in the realm of potentiality and the superposition of images and meanings in dreams.[19] Bohm points to the experience of music, in which the power of the music for us depends on the interaction of each newly heard note with an immediately felt sense of all the notes that have gone before. Are we here touching on further dimensions in the implicate order?

Comfort suggests quite explicitly that this may be so; that in superposing dream images we may be looking down another dimension, in which dreams are co-present and separate, just as when we look down a straight line of trees they appear superposed. And Bohm suggests that the power of music comes from the connectedness of the notes in an implicate order which we directly perceive. That is to say, we are perceiving "the relationship of co-present elements, rather than the relationship of elements that exist, to others that no longer exist."

The Representation is the Represented

We could summarize the various interpretations that we have described so far in the following way: For Bohr, what happens is what *is* observed and is determined by *how* it is observed. Beyond this, "reality" has no further description. When other investigators try to describe the unmanifest reality, they do so in terms of a realm which is not consciously known. This realm is variously described as consisting of potentia (Heisenberg), propensities (Popper), implicate orders (Bohm), or other worlds (Everett). In each of these descriptions, that this particular world is this actual world, observed and lived by someone, is related to something vaguely called the "consciousness" of that someone.

All these interpretations of quantum mechanics, including in some sense the Copenhagen interpretation, seem to imply a connection between actuality and someone's consciousness of it. But as we will see in Chapters 12 and 13 in what is in fact one of the main themes of this book, the idea of a continuous, unitary "consciousness" belonging to each observer is so open to question at this point as to be almost worthless. This seems to be a crucial point, not only for each of us personally, but also for each of the interpretations of quantum mechanics. For the Copenhagen interpretation, certainly only what we observe is real, and there is no meaning to reality beyond that; then why do we all seem to have the same "reality"? For Wigner, whose minding is it that brings potentiality into reality? For the Everett model, your minding of it is all that distinguishes this world from all the other real worlds in which replicas of you exist. In speaking of the anthropic principle, Davies suggests that it may be our presence in this world that selected this one. But what is meant by "our presence"— presence of the organic molecules making up our bodies, presence of a perceiving organism, presence of "our consciousness"? Clearly, he means all of these, yet it is only the latter, "consciousness" of ourselves as "observers," that is in *this* particular world. But if your consciousness is totally separate from mine, each isolated within our bodies, then how is it that our presence has brought about such seemingly similar or at least overlapping worlds? For Bohm, consciousness and appearance (explicate order) arise together as they do for Bohr. They arise out of a deeper implicate order in which the separation of individual "minds" or "consciousnesses" is questionable. And at this level we can no more say that "consciousness" is prior to "matter" than that "matter" is prior to "consciousness."

Some of the problems quantum physicists encounter in trying to interpret the theory seem to have arisen from the classical assumption that consciousness belongs to a separate mind which has a particular identity. And certainly the word *consciousness* is used very loosely. It would appear that the next step in trying to understand the relation between manifestation and consciousness would be to study the biological process of thought and perception or, through meditation training, to study directly the processes of mind.

Perhaps we can turn again to Niels Bohr: "The idea of complementarity is suited to characterize the situation, which bears a profound analogy to the general difficulty in the formation of human ideas, inherent in the distinction between subject and object. In particular the contrast betweeen the continuous onward flow of associative thinking and the preservation of the unity of the personality exhibits a suggestive analogy with the relation between the wave description of the motions of material particles, governed by the superposition principle, and their indestructible individuality." Bohr was not aware of the details of the process of perception, as we now are, but those details as they unfold seem to exemplify his statement more and more.

This brings us again to Renouvier's philosophy which was one of Bohr's sources of inspiration: "The represented *is* the representation," and beyond this there is no form, no conceptualization. Holographs, implicate orders, other worlds, all are ways of trying to point to the unconditioned which cannot be finally objectified by concept and form.

In Part Two, I have shown that our "commonsense"—that is to say, conventional, unexamined and untrained perception of our world—may be strongly influenced and narrowed by unconscious presuppositions received in part from nineteenth-century physics and biology. I also showed that the physics and biology of the past half century have seriously undermined these limiting presuppositions.

Although we cannot jump directly from these extended views of time, space, and matter, which are still highly abstract, to our immediate experience, we *can* say without question that the impoverished sense of the world provided in the classical view has been broadened. We are free to return to the richness and fullness of our experience, to rediscover for ourselves the nature of space and time, and to appreciate the natural energetic order in our world. Prigogine has begun to show how an understanding of change, of *becoming*, might be introduced into physics on a primary level to extend the domain of classical and relativity physics, which was a static, timeless physics of *being*. The result is, as we have seen, a view of evolving formation in space-time. We have seen that categorization of the world into "things" that are living and "things" that are not is also suspect. Our studies of quantum mechanics have suggested that the unconscious categorization of our experience as perception of an "external world" of separate material objects is highly questionable. We might ask, is it possible to go back even deeper in our questioning of our presuppositions? Is it possible to construct an understanding of the actual lived world in which even space, time and the division between an internal and external world are *not* presupposed?

In order to understand this question at all, it is necessary to look more deeply at our immediate experience. We experience the world through the body, *with* the body. We do not experience an abstract world of space, time and objects; rather we see *with* our eyes, hear with our ears, and so on. A more complete understanding of the world must take into account this presence of the body that accompanies any particular content or organization to our perception.

This is the point of view of process philosophy, which we will study in Chapter 14. The tremendous power of process philosophy lies in the fact that it provides a context for a broad understanding of the world which can include the sciences as well as poetic and religious insights, yet which is based not on high level abstraction, but on discovering the profundity of immediate experience. In Part Three we will discuss, as well as process philosophy, the stages of perception, according to Vajrayana Buddhist meditative insight. These two descriptions of the process of perception, as experience arises moment by moment from the unconditioned ground, are strikingly similar, in spite of the quite different cultural and philosophical backgrounds in which they evolved, and the different languages in which they are expressed. Yet the similarity may not be, after all, so surprising, since at this level we are dealing with observation of the basic processes of the human mind/body which precede cultural differences. We will, however, begin by reviewing a few of the discoveries and hypotheses in cognition and neurophysiology which, in the past decade or so, have begun to shed light on the details of perception. These discoveries suggest that the process of perception is nothing at all like the naive camera theory. Rather they seem to be pointing in a direction close to that of Whitehead and of Vajrayana Buddhism, namely that each moment of perception or experience is self-created afresh. Perception is a moment-by-moment process in which a world of experience arises along with the interface between "organism" and "environment," and the notion of an "external world," or for that matter of an "internal world," dissolves.

Part Three
Fruition:
Perceiving
Ordinary
Magic

12 Flickering Perception

We will begin our discussion of perception by looking at a few suggestions from cognitive and neuropsychology which throw serious doubt on the standard "representation" or "camera" theory of perception. This theory suggests that the end point of the perceptual chain is a more or less accurate representation in our brain of the world outside us. According to this view, "consciousness" observes this representation, or in a more sophisticated version, the representation is somehow self-observing. Rather like looking at a photograph, we thus gain a more or less true notion of the world. That it must be more, rather than less, true, is deduced from the fact that the organism is able to function in the world. In some vague way this is what most of us believe about how we perceive the world. We tend to think that our perception is, in itself, pure, and that all interpretations and emotional coloration are secondary to this pure perception. In this chapter we will see that matters are not quite so simple. Interpretation seems to be the very nature of self-conscious perception. Furthermore, perception appears to be more like a flickering movie with gaps between the frames than it is like a still photograph, the frames being as much a creation or projection of the observer as they are of the "outside world."

In order to understand our changing experience and how conceptualization as well as mindfulness and awareness might enter into that experience, we need to look very closely at the details of perception, particularly at the moments of transition or branch points between perceptions. Let us then consider the example of visual perception. We will trace, in its major outlines, what happens to an image, perhaps the face of our friend, after it strikes the retina.[1]

Interacting Subsystems of the Brain

But before following the pathway of a visual input we must first outline the major subsystems of the brain through which such a visual input must wend its way. The most clearly worked out of the subsystems are a vertical division into three parts, and a lateral, left-right division. The vertical division is known as the triune brain and was first pointed out by Paul Maclean following earlier leads of Papez.[2] The three divisions seem to be quite distinct from one another and to have been superimposed on one another in three distinct evolutionary stages. The first brain, known as the reptilian brain, is similar to the brain found in prehistoric as well as present-day reptiles. The second brain, the "old mammalian" brain, is similar to that of lower mammals—rats, rabbits, kangaroos, horses, and so on. Third, there is the "new mammalian" brain, especially developed in primates and humans.

Maclean proposes that there is a kind of "neural chassis" subordinate to all these three brains, composed of the lower brain stem, the spinal chord, and the midbrain. Without the three brains as three almost independent "drivers," this neural chassis would be like an unguided missile, a runaway car without a driver. The neural chassis itself coordinates aspects of self-preservation, such as breathing, blood circulation, blood pressure, digestion, movement, and the selection of environmental stimuli essential for self-preservation.

The first brain, the "reptilian" brain, coordinates a rich set of behavioral patterns: territoriality, ritual fights, formation of social hierarchies, greeting, ritual courtship, migration, hoarding, and playing.

The second brain, the "old mammalian" brain or limbic system, receives information from both the "outside" world and the "inner" world of the organism and contributes significantly to the formation of personal identity. The limbic system registers basic affects such as hunger, thirst, pain, repugnance, and general motivating patterns of behavior: searching, aggression, caressing. It is concerned with attention, emotion, learning, and memories. It mediates messages received from the outer environment, "suffusing these with moods ranging from rose-coloured anticipation to dark disappointment, as when an anxious mother meeting a train sees her son's resemblance in every passing boy." Areas have tentatively been identified mediating a range of basic moods: rage-fear; fight-flight; pleasure-pain; expectation-actuality; tension-relaxation. The first two brains are not "unconscious"—their manifestation can be recognized in a rich array of bodily postures, facial gestures, and behavioral patterns. However, they cannot verbalize their responses and cannot be direct objects of "self-consciousness."

The third brain, the neocortex, is according to Maclean comparable to an immense neural screen capable of abstraction, symbol formation, language, and logic, and of forming interconnections between the senses. It is the termination of sensory inputs and is involved in control of motor activity.

The brain also appears to be to some extent laterally divided in function between the two hemispheres. The dominant hemisphere, in which necessary structural parts for speech and language ability are located, appears to function in a more literal, analytic way. The opposite hemisphere (usually the right) has a more holistic, intuitive mode. The left hemisphere is also apparently an important structure for self-consciousness, which has led many people to associate it with language ability. These two different modes of functioning were initially discovered when, in an attempt to cure extreme epilepsy, patients' brains were split and the millions of fibers joining the two hemispheres surgically severed. However, subsequently methods have been developed for detecting the left-right difference in normal subjects.

We can summarize the tendencies toward different modes of functioning in the two hemispheres:

Left Hemisphere	*Right Hemisphere*
Verbal	Nonverbal
Analytic	Holistic
Reductive	Synthetic
Linear-sequential	Visuospatial
Convergent thinking	Divergent thinking
Self-conscious	Not self-conscious

Hampden-Turner comments: "The discovery of the significance of the two hemispheres is clearly of momentous importance. The question as to whether introspection and insight into the nature of the human mind is worthwhile, or even possible, appears to have been emphatically confirmed. For the things that physiological investigation has now discovered the minds of men have long intuited."[3] Unfortunately, there has been a tendency to overestimate the importance of this discovery and promote it as the one major factor in human evolution. Localization of functions in particular areas of the brain is itself questionable. What this research does show is that such functions as those listed under "Left Hemisphere" and "Right Hemisphere" are functionally grouped together.

We should mention also the importance of the reticular activating system (RAS), a network of cells which lead from the spinal cord through the reptilian brain to the thalamus, a kind of relay station through which sensory messages pass on their way to the cortex. Fibers from the RAS also pass to the hypothalamus, which is involved in emotion and drives. The RAS is the "alarm bell of the brain." It is responsible for alertness, for awakening the brain to activity. When the RAS is damaged, an animal becomes irreversibly comatose.

The cortex is divided front to back by a large fold, and the somatosensory and movement areas are localized on either side of this, almost as a projection of the body. Areas for reception and transmission of impulses to the other senses, as well as, in the left hemisphere, areas controlling speech, have been mapped. There are also millions of cross-links between all these areas as well as millions of connections down to the lower brain. In particular, there are connections from the various early sensory areas to the limbic system and from there to the cortex, especially the prefrontal areas involved in planning and the feeling of meaningfulness.

Complex Interconnections in the Visual System

The first stage in visual perception is the fragmentation of the image on the retina into the individual, connected responses of ten million cones and one hundred million rods. The final stage is a conscious perception. How then is such a picture reconstructed (or perhaps just constructed)? The "reconstruction" begins in the nervous system of the retina itself, which already begins to abstract patterns or features from the mosaic of the retinal image. These are passed on down the optic nerve. The optic nerves from each eye cross and half of the retinal image from each eye passes to each hemisphere of the brain. The right half of the visual field from each retina is connected eventually to the left hemisphere of the cortex, and vice versa.

Before reaching the cortex, the millions of fibers from each side of the optic nerve enter a part of the thalamus known as the lateral geniculate nucleus (LGN). For every one fiber entering the LGN from the retina, more than eighty enter the same point from other areas of the brain, including connections returning from the cortex. There appears to be some kind of reconstruction of an image on a layer of neurons in the LGN, which for this reason has been called the "internal retina." The several million fibers carrying the visual message pass on from the LGN to several separate areas of the cortex. Altogether these cortical areas contain several hundred million neurons, that is, several hundred times more than entering neurons from the LGN. Leaving aside whatever processing occurred to the visual image in the LGN, these messages are now entering a subsystem several hundred times larger.

The visual areas of the cortex connect to the prefrontal lobes. These lobes have been called the "interpretive lobes" by Wilder Penfield, since they appear to be involved in interpretation of the relation of the individual with his immediate environment: interpretations such as "familiar," "frightening," "closing in," or "moving away." The prefrontal lobes are also connected to the limbic system, the seat of the emotions. So, as Eccles says, "One can think of the prefrontal cortex as being the area where all emotive information is synthesized with somoaesthetic, visual and auditory to give conscious experiences to the subject and guidance to appropriate behavior." And again, "by their connection to the prefrontal lobes, the hypothalamus and limbic systems modify and color with emotion the conscious perceptions derived from sensory inputs and superimpose on them motivational forces." Furthermore, the visual areas of the cortex also connect *directly* to the limbic system, as do the prefrontal lobes, so complicated feedback or feedforward loops may be built up. Finally, the visual cortex itself connects back to the LGN, so that the visual messages entering the LGN may be modified at this relatively early stage by this already richly emotive interpretation. The LGN is also connected to areas responsible for visual recognition and for movement of the eyes to objects of interest, so that the feedback from the cortex is able to influence even what enters the eye at all.

There is one further, very important connection that we must mention—that of speech, and the ability to form concepts or names. It appears that one of the main areas known to be necessary for language (the Werniche speech area) is also the area used for association between the senses (cross-modal). To quote Geschwind, one of the experts on these areas, "The ability to acquire speech has as a prerequisite the ability to form cross-modal [sensory] associations. In sub-human forms the only readily established associations are those between non-limbic [visual, tactile or audial] stimulus and a limbic [emotional] stimulus. It is only in man that associations between two [sensory] stimuli are readily formed and it is this ability which underlies the learning of names of objects."[4] This cross-association, mixing of inputs from various senses, as well as language, occurs in a cortical region which again feeds into the LGN—the primary switchboard for visual images.

To summarize: although I have given the barest sketch, we can easily see that the visual system is a complex network of interlocking loops connecting what we see with our other senses, our emotions, our interpretations, our interests, and our name formation, and all of this feeds back to the original point of entry of visual information from the optic nerve. How could we ever have thought that the eye was like a camera taking its pure unadulterated little snapshots of the world "out there"? And of course similar systems exist for the senses of hearing and touch. The sense of smell is different in that its connections are primarily to the limbic system and only later to the cortex, no doubt partially accounting for the particular connection of the sense of smell with strong emotions.

I have said nothing about the final reconstruction into a unified "conscious" image, because almost nothing is known. There is some indication that some neurons in the visual cortex respond to particular, very simple features in the environment: length, thickness, and brightness of lines and their orientation and other simple geometric features. Beyond this there is indication that some neurons may be very specific—for example, one neuron, in an experiment on monkeys, appeared to be fired specifically by the silhouette of a monkey's hand. But how all this is assembled is unknown from the standpoint of neurophysiology.

If we adopt the naive commonsense view—namely, that there is a real, solid world of "things" outside of the body, that our "consciousness" is localized inside the body, and that somehow the brain through all these complex pathways is able to construct an image, a "representation," of the outside world—then we must certainly question how "true" such an image could possibly be.

The typical view of neurophysiology is summed up thus by Vernon Mountcastle, considered by many the dean of psychobiology: "Each of us believes himself to live directly within the world that surrounds him, to sense objects and events precisely, and to live in real current time. I assert that these are perceptional illusions. Each of us lives within the universe—the prison—of his own brain. Projecting from it are millions of fragile sensory nerve fibres in groups uniquely adapted to sample the energetic states of the world about us: heat, light, force and chemical composition. That is all we know of it directly, all else is logical inference, sensation is an abstraction, not a replication of the real world."[5] Mountcastle's statement depends on the assumption that each of us *does* live within his brain, and that there *is* an "external," "real" world around us. In the next chapter we will find good reason to doubt this assumption.

Conjectures and Perceptual Matching

Let us turn now to direct studies of perception. The work of Richard L. Gregory, professor of neuropsychology and head of the Brain and Perception Laboratory at the University of Bristol, is of some significance.[6] Gregory has spent over thirty years studying perception and in particular visual perception. He has made a special study of visual *illusion*. Gregory realized that perception is extremely close to our thoughts, ideas, and hopes about the world. It would therefore be difficult to tell, if we simply study the normal functioning of perception, to what extent our theories about it are influencing our studies. We need, therefore, to look for a situation in which nature speaks to us almost in spite of ourselves. The best way to find out about perception beyond our presuppositions about it is to look at cases when it seems to go wrong: cases of visual illusion.

Gregory describes four general kinds of visual illusion: distortion, ambiguities, paradox, and fiction. An example of *distortion* is the well-known child's puzzle in which we ask whether two horizontal lines are the same length (see Figure 4). The answer is almost invariably that the top line is longer. And even when we know that the two lines are in fact the same length, we still continue to see the top line

4.

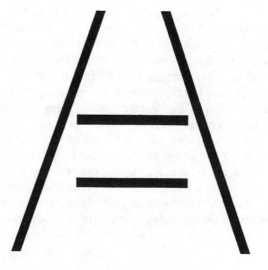

as longer. This is a simple example of the phenomenon of size constancy. For example, if you hold your thumb a few inches in front of you and then move it away, it is usually perceived as remaining roughly constant in size. Gregory proposes that, in regard to Figure 4, the brain is used to seeing, in the actual world or in pictures, many instances of converging perspective lines and knows that in those actual situations the horizontal lines would not be of equal length; therefore they are perceived as unequal.

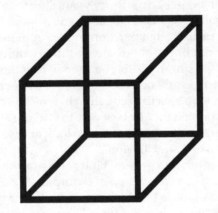

5.

An example of *ambiguity* is another famous illusion: the Necker cube (Figure 5). We perceive such a cube as three-dimensional. However, point A can be perceived as a corner close to us, or in the back. The cube flips back and forth between the two. Furthermore, we can make it flip back and forth with some kind of inner effort, or, if we relax, it seems to flip back and forth itself. We can show that this flipping has nothing to do with eye movements. For example, we can look at the image under a bright light and then watch the afterglow—the retinal image. This too flips. Therefore, the flipping must be taking place at a deeper level of the visual system. Gregory proposes that some higher brain function is flipping between two possible conjectures as to which way around the cube "is," that is to say, "is" if it were a real wire cube in the physical world. If it were such a wire cube, it would *have* to be one way or the other, and the brain seems to be preprogrammed to try to resolve such ambiguities by choosing between conjectures. Another example of this flipping is the young lady/old grump (orginally called the mistress-wife illusion) (Figure 6).

This also has a disturbing quality and is slightly closer to showing us that such illusions might happen in visually quite complex situations in the real world.

6.

An example of *paradox* is seen in Figure 7—a flat figure which looks three-dimensional but which, if we follow the edges, we know cannot be three-dimensional. Escher has made brilliant use of this kind of paradox. We are startled and sometimes quite irritated by his work because it does not fit any possible conjecture as to what is "really" going on in the scene.

7.

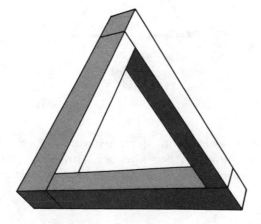

8.

Finally there is the *fiction* illusion, of which we see an example in Figure 8. We actually *see* a white, solid triangle joining three points. It is not just an intellectual process. Even when we try to think that there is really no white triangle there, we still see it. This is even more striking in Figure 9, where the black angle lines are not cooperating with what we see—we see a triangle with curved, concave sides.

9.

An example of such filling in more natural circumstances is the spotted dog in Figure 10. Again Gregory is proposing here that we are making a conjecture based on seeing that dog's ears and collar. We fill in the rest of the dog based on that conjecture.

All of this is pointing in the direction that certain features are arising in the interaction between the visual system and the environment and that the higher, conceptual mechanisms of the brain are filling in the rest by a process of matching the most likely hypothesis.

10.

This theory is further confirmed from two directions. First, as we saw in our description of the brain mechanisms in vision, although there are very few known feature-detecting neurons in the brain, there are some. And these all have to do with very basic aspects of the environment—brightness, geometric patterns, perspective, sudden movement, and so on. Second, another psychologist with a large following, J. J. Gibson, points out that there are certain invariances in the light which arrives at any point in the real world. For example, as we walk along in a field, the speckled patterns of the grasses seem to move toward us, getting larger and moving to each side of us. It is by the continuing interaction with these moving perspectives and patterns that we guide our own movement. When we actually examine the nature of Gibson's invariances, we find they are quite abstract and could very well correspond to some of the neuronal feature detectors: the faces and edges of objects, the texture of the ground as it grows more dense in the distance, and so on. Some of the examples Gibson gives of the way these invariances, or "affordances" as he calls them, in direct perception seem to confirm rather than contradict Gregory's view:

"A rigid object with a sharp dihedral angle, an edge, affords cutting and scraping. It is a *knife*.

"A graspable rigid object of moderate size and weight affords throwing. It may be a missile or only an object for play, a *ball*."[7]

We see here a very clear description of the process of guessing based on ambiguous features of the environment. All of these arguments apply equally to the other senses: we fill in words we do not hear—for example, if an after dinner beverage is being served and our hostess says, "Would you like tea or (Clang)?" we fill this in as "Would you like tea or coffee?" Audial *ambiguities* are also quite common—for example, if a student is visiting a science professor in his office and in the middle of the conversation the professor stands up and says, "Well, it's ten o'clock, I really have to go to the laboratory now," the student might hear this as "Well, it's ten o'clock, I really have to go to the lavatory now," and wonder why the professor felt it necessary to explain in such detail. An example of a tactile ambiguity might be when we feel a sharp prick under our shirt and leap out of our chair, momentarily taking it to be a wasp.

Fear and the Central Conjecture of "I"

Now, all these are instances in which perception is not quite functioning, but they seem to point directly to the mechanisms of perception even in its normal mode: as we move through the world, certain features arise about which are made conjectures to find the best fit. In the normal course of the movement of the organism through its natural world, perception works well; we are usually able, more rapidly than we can notice, to find a conceptual match. We might ask, however, what happens when the brain *cannnot* find a conceptual match to perceptual features? Richard Gregory's work suggests that consciousness of our perceptions may be, precisely, a biological response to perceptual mismatch. And when we look deeper into this, we find that the motivating energy for such consciousness may well be fear.

Behavioral biologists have examined the general parameters of our behavior, the emotional energies that are based on the interaction between nervous and endocrine systems. Rage, pain, fear, pleasure, lust, attachment, depression, and greed: each one of these has a biological and endocrine basis in these systems. Of particular interest in this analysis is the interconnection between these basic energies, the central role of fear and the mechanism of fear itself. That fear is very close to rage can be seen in the outward expressions of both: the facial expressions of fear shade imperceptibly into those of rage; the fight-flight response is a delicate balance and seems to be based on a single range of internal responses; the structural parts of the brain affecting fear and rage seem to be to some extent the same. Pleasure also, at this very basic level of biological energies, can be understood as the release of stress, that is, the relaxation of a temporarily fearful situation. Attachment also seems to be intimately connected with fear, the fear of loss: "Fear of being separated unwillingly from an attachment figure is an instructive response to one of the naturally occurring clues to an increased risk of danger," reports John Bowlby, one of the great psychobiologists, author of a three-volume work on attachment and loss.[8]

It is the mechanism of fear itself that is most interesting. There is strong confirmation from many different sources of a suggestion first put forward by Donald Hebb in the 1940s that *fear is the result of perceptual mismatch*.[10] That is to say: the fear reaction is aroused whenever the organism has a perception which does not match its anticipation of what it should perceive in a particular situation. To take one of the simplest examples, if chicks were made to experience repeatedly the shadow of a goose passing over their nest, they crouched (a fear response) when the shadow of a *hawk* passed over. However, if they were habituated to the shadow of a hawk, then they crouched when the shadow of a goose passed over. Many experiments done in different circumstances on animals including chimpanzees and humans make this hypothesis very likely. It is also understandable in terms of evolutionary theory: it is much more efficient to wire into the nervous system a genetic tendency to fear when confronted with any stimulus that does not match our expectation, than it would be to wire in individual fear reactions to all the multitude of things which might be threatening to a particular animal in its normal life situation. What is the implication?

When the conjecturing brain is not able fast enough to find a conceptual match to the perceptual features arising in our fields of vision, hearing, etc., the result is fear. Fear then becomes the basic guidance of our perception. We know that we have made a good match when there is absence of fear. Fear keeps us on our habitual track as we go through the world. Now, all this makes very good sense in the case of an organism moving through its natural surroundings, recognizing objects like rocks and trees and other members of its species in this way. But when we think that most of human life is not spent recognizing rocks and trees but trying to recognize human facial expressions and language with all its subtle intonations, then it becomes a different matter. As behavioral biologist Melvin Konner says: "No reasonable analysis of human behavior can fail to grant that situations that most seriously jeopardize human survival and human dignity past, present, future owe much more to irrational fear than to irrational rage."[10] But it now appears that most perceptual situations, when taken for granted, *are* based on fear. We realize that trying to keep our perceptions constantly matching our conjectures about what is going on is a very tricky business, in which the unfamiliar is always lurking very close by. We usually cope by narrowly circumscribing our lives so that we do not come across the unfamiliar too often. Alternatively, perhaps we could say that each moment is in actuality unfamiliar but, to avoid fear, is made familiar by our habitual conjectures. That is to say, our lives, when we do not recognize the perception process for what it is, are ringed by a circle of fear. And for this very reason we try, constantly, to avoid seeing the basis on which we live our lives. If, on the other hand, we do recognize the fear and experience it rather than try to avoid it, then we are not forced to accept the habitual conjecture. Unfamiliarity then can be a source of possibilities.

Now, what could be the basis of these conjectures that our higher brain functions are continually making? Presumably this whole process has developed as a means both to warn the individual of danger and to enable him or her to predict and manipulate the environment more effectively. It would certainly seem that these are the primary purposes of the conjectural aspects of perception. Therefore, it would seem to follow that a very large group of conjectures would be oriented from the point of view of the existence of that individual as a separate entity: the conjecture of an "I." We could call this conjecture of an "I" a central conjecture, around which many other conjectures are organized. It would appear that this central conjecture of an "I" is a major organizing conjecture both individually and culturally. As we discussed in Chapter 7, even at the biological level, it seems likely that consciousness of self and other as separate biological entities, and of one's inner sensation of bodily states, evolved as a means of imaginatively empathizing with other individuals in order to predict their behavior.

Let us now try to go deeper into this process of constructing an "outside world." So far we have seen that the organism receives a message of some set of features arising in its interaction with the world and forms a hypothesis concerning what is "out there." We must remember that in selecting a particular set of features the organism is taking for granted other features that form the background or context for that particular perception, in particular features of his own body.

Perception Flickers

Perception is a process that takes time. It has been shown that it takes approximately one half second for a sensory stimulus to become conscious. In comparison, it takes approximately one-hundredth of a second for a sensory stimulus to reach the cortex and one-tenth of a second for an "automatic" prelearned reaction to a stimulus. This was demonstrated by Benjamin Libet afor the sense of touch, whose pathways in the brain are very closely similar to and even interconnected with those of the sense of sight. Libet performed experiments on patients undergoing brain surgery who were conscious and consented to take part in them. He showed that direct prolonged stimulation of the touch area of the cortex caused a sensation on a part of the skin as if it were being pressed. Using this finding in some very ingenious experiments in which he compared the responses of such cortical stimulus with direct pressure to the skin, Libet found the following results:

1. A single, brief impulse applied to the skin can be detected consciously.

2. A single, brief impulse applied directly to the cortex is not detected consciously.

3. A continuous train of impulses (1/20 second apart lasting for less than 1/1,000 second) has to be applied directly to the cortex for *at least one half second* for these to be detected consciously.

4. A single, brief impulse to the skin was not detected consciously if the cortex was stimulated within one half second after the skin stimulation.

Libet concluded that, although a single impulse to the skin could reach the cortex in less than one-hundredth of a second, this impulse had to be somehow repeated and elaborated at the cortex for at least one half second before it was consciously detected.

Libet also showed that this delay occurs in relation to perception not only of the outside world, but also of our own muscular movements. He asked subjects to move a finger quickly at any time they wished and showed that EEG patterns corresponding to beginning the move occurred at least one-third second before the subject thought he wanted to move.[11]

Such time delays between action and consciousness of that action have been corroborated in many experiments and observations which implicate the frontal cortex in planning, attention, and projection of meaning and future orientation. And it has been shown that this frontal cortex acts on the perceptional process before this process reaches "consciousness." That is, our perceptions are already tremendously digested and programed before we normally become conscious of them. A typical such experiment, by Karl Pribham, involved monkeys choosing between a card with a circle or a stripe on it by pushing an appropriate bar and receiving a reward. The monkey's brain waves showed that at a very early stage in the processing of the visual information, his brain had already selected the correct bar to push. He was clearly *intending* to push the bar a comparatively long time before he actually did so. For another example, slamming on the brakes when someone cuts across in front of our car takes only one-tenth of a second, but it takes at least one half second for us to realize what has happened. We are probably all familiar with this delayed reaction. We might wonder why we have the illusion that we experience in present time, when we are in fact experiencing the world after a one-half-second delay (a very long time). Libet suggests that the brain ingeniously refers its conscious impression to that first arrival at the cortex (which was only one-hundredth second late) and projects its experience back to that moment, thus giving the impression that experience coincides with reality.

We might note here that experiments and long clinical experience with biofeedback have shown that it is possible for a subject to become conscious of, or to influence without becoming conscious of, fine details of internal bodily events normally far below the threshold of consciousness. Pulse rate, blood pressure, types of electrical activity in the brain (brain waves), all can be brought under the control of a subject in biofeedback. And this is precisely what is said of meditation: that we can train our awareness to be attentive to processes briefer than one-hundredth of a second. Traditionally it is said that moments of awareness can be detected, in meditation, of a duration of one-sixtieth of the snap of a finger. By this is presumably meant one-sixtieth of the shortest duration detectable by untrained perception, i.e., approximately one-sixtieth of one-tenth of a second. Therefore, it is possible through training to experience the whole process of perception.

Now, we know that a single image can reach the cortex in one-hundredth of a second. Does the organism form its final decision concerning the object in this time? Probably not. We saw in fact that it takes one half second to become conscious of an image. And other experiments have shown that images flashed for only one-hundredth of a second are not consciously remembered by a subject, although they can be recalled in dreams or hypnosis.

There are many possible explanations of Libet's experiments, which are still quite crude. However, aided by the insights of the meditative tradition, I suggest that what is possibly happening in the one half second it takes for an image to become conscious is that there is a flicking back and forth between nondual perception and conceptual conjectures as to what is "outside." Gregory Bateson described a very similar zigzag, stepwise process of perception in which the flickering is between process and form, between news of differences in the world and hierarchical categorizing of this news. "A world of sense, organization and communication is not conceivable without discontinuity."

We cannot from these observations say what lies in the gaps between conjectures. But we can suggest that the driving force of the conjecturing process is the need to match perception with expectation, to avoid fear, and to continue to exist in a manageable world. Possibly the gap is an open field of perception in which there are vastly more possibilities, many of them quite *un*expected, than the one finally settled on. We can certainly say, however, that perception appears to be a far more dynamic, interpretive, and participatory process than merely reading a photograph-like image.

We must, finally, raise two further, related questions, namely, What is it that is doing the flickering? And what is it that finally settles on the "best" conjecture? To answer these extremely subtle questions will certainly require more than the kind of crude experimentation we have presented so far. However, we can suggest that to speak of a unitary and separate "consciousness" of the individual organism is far too vague. The process we have described is clearly not taking place at a level of ordinary consciousness—although, as I have indicated, it is said to be possible to become aware of the process through training. To see that we are not discussing ordinary "consciousness," consider the following scenarios—typical everyday experiences, in some form or another, for most of us, experiences of "absentmindedness":

1. You are sitting on a chair in the kitchen with a friend. You are very absorbed in a book. You reach up and turn off the tap, then continue reading. Later your friend says, "I am glad you turned the tap off, the dripping was bothering me too." You have no idea what he is talking about. You did not hear the dripping, and you remember only that you were not aware of reaching up to turn it off.

2. You are sitting on a chair in the kitchen with a friend. You are very absorbed in a book. You notice a noise, realize it is the tap dripping, and turn it off. You then resume reading feeling glad the noise has stopped.

Perhaps it would be more accurate to say "minding" than "being aware" or "conscious." The double meaning of the English word "mind" is perhaps not accidental. "Did you mind the dripping tap?" "No, I did not mind it. I did not even notice it." Or, "Yes, I minded it so much I got up to turn it off." We could almost say that the world consists of all that we mind—whatever is irritating or attractive enough for us to notice it, to mind it, is included in our world, while the rest is not. This minding goes deeper than the merely conscious level: in the second case you had to mind the dripping tap before you could notice it and turn it off. In the first case, although you may not consciously remember the event, it could be recalled in hypnosis, or in the extraordinary experiments of Wilder Penfield.[12] While conducting surgery on epileptics, Penfield electrically stimulated various areas of the cortex. His patients were conscious and able to report what they were experiencing. Penfield was able to cause his subjects to relive memories in tremendous detail. They reported that they felt they were *actually experiencing* these memories—reliving them, "minding" them. At the same time, they were able to experience themselves in the operating room. There appeared to be two distinct simultaneous experiences. This, again, raises the question of who, if anyone, we are referring to when we speak of the "consciousness," a question that arises again in the phenomenon known as "blindsight." Human subjects who report being unable to see anything at all in a certain region of the visual field, being totally "blind," nevertheless if asked to "guess" the location of particular objects do so with accuracy far above chance.[13] Although the conscious subject is not "seeing," the organism is visually detecting the location of objects. Again we ask, who, or what, is "seeing"? As we will see in the next chapter, this question is closely connected with the question: What is seen, i.e., what is the "outside world"?

13 "Outside World" as Conjecture

We must now turn to consider the role of the "outside world" in perception. The traditional assumption of the division of nature into an organism which is perceiving and an "outside" which it perceives is an arbitrary, although useful, division which the organism makes in order to survive. It is an aspect of the "I" conjecture. But it *is* arbitrary; it is by no means an absolute division of nature, or one on which we should base a broader view of perception, as we will see in this chapter.

The Nervous System as an Autonomous Unit

In the previous chapter we saw that the nervous system, particularly the brain, is a tremendously complex system that has *interconnections* at all levels. It seems to function as a unit. Any change in any subsystem of the brain results in adjustments thoughout the entire system.

Based on these considerations, Varela has suggested that it is more useful, and appropriate to the organism, to speak of the nervous system as an autonomous unit, an operationally closed system. This system passes through its own internal states (its conjectures), for which interactions with the "external world" act as triggers. We do not "know" the external world; we know our internal states, or perhaps we should say that the internal states know themselves. But these are in a constant state of change, and that change is triggered by "interactions" which we refer to as the "external world" or "reality." To help us visualize the sense in which the organism passes through its own internal states, Maturana and Varela use the analogy of an airplane pilot in a terrible storm with zero visibility trying to land his plane with instruments. To the pilot there is no difference at all between this situation and a simulated storm in the safety of the training cabin. When the plane lands and everyone congratulates him for making such a beautiful landing in the fierce storm, he might say, "It felt just like the hundred practice landings I did." He was responding to the internal states of the plane, which were triggered by messages from the airport control tower, but he was not responding to the storm itself. As an analogy for the human interaction with its environment, a plane landing on automatic pilot might be better: we are not assuming that the human has a "little man" inside him, separate from the body/mind and guiding it, i.e., we are not assuming a soul, or a separate "mind." The various states that the airplane can go through in response to the messages from the control tower, such as angles of wing elevations and so on, are a function of the internal structure of the plane itself.[1]

Like the automatic pilot, the nervous system, Varela suggests, is responding to its changing environment by passing through states which are constrained by its own structure as well as the history of previous pathways of states it has passed through in previous interaction with the environment. The important point is that from the point of view of the nervous system its available states are not by any means a direct mirror of the "outside world." As Varela says, "[This research] means moving away from viewing the brain as a device which takes input in the form of information to act on. Rather it means moving towards viewing the brain as a system characterized, not by its inputs, but by the operational closure of its dynamics of states, defined as a relative balance of activity between neural surfaces in a manner such that every change of state of the system can only lead to another change of state of the system itself . . . . For us, the origin of knowledge and the making of sense, does not resemble the design of a system which is optimized to match a given external standard [like a camera taking snapshots of the 'outside world']. We could say it resembles, rather, a tinkering, a dynamic sculpting, a building of structures from the materials available to an organism that it puts together as they appear in a drift which follows one of many possible paths."

Varela points out that this point of view applies not merely to some nervous system we are studying, but also to our own nervous system itself. He concludes, "If we are right, our human life, our experience right now, is but one of the many possible creodes [pathways] of knowledge, where the immense background of our biological structure and social practices is inseparable from the regularity we discern in both world and self. When we follow this logic all the way through, we can understand the world in which we find ourselves as neither separate nor distant. But also, as one where we have no fixed reference points left."

Criteria For Mental Process

Let us look further at how mind or mental process might arise in this dynamic process of interplay of an organism with its environment. To do this I would like to discuss the view of Gregory Bateson which also takes us to a much deeper level of explanation where, as we shall see, the absolute distinction between nature and mental process is questioned. Gregory Bateson worked during the early part of his life, in the early 1930s, observing and participating in the cultures of New Guinea and Bali with Margaret Mead. In the 1950s he worked as a psychiatrist and began to apply the newly developing insight of information and communication theory. During this phase, he developed the double-bind theory of schizophrenia, and also continued his lifelong exploration of social and cultural behavior in humans and animals. Toward the end of his life, this wise and gentle man was mentor to a large number of young scientists working in many fields: neuroscience, cognitive psychology, ethology, information theory. He was elder to many of the young people living through the confusing upheavals of the 1960s and 1970s.

Just before he died in 1980, Bateson finished *Mind and Nature: A Necessary Unity*, in which he offered us the fruition of more than fifty years of participation and observation of the life and behavior of animals, including humans.[2] Bateson's main thesis is that the patterns in mind are a reflection of the patterns in nature, and that it is these patterns and only these patterns that we can "know," in the ordinary sense of "have direct experience of." We cannot "know" the things-in-themselves, nor can they know each other. Therefore, the regularities in these patterns are the closest we can come to "ultimate truth."

Bateson proposed a detailed set of criteria as a response to the question: of all the forms of life, learning, and mental process which we know, what is the pattern which connects them?

1. A mind or any living system is composite—it has interacting parts—and only when we look at a system as an aggregate of such parts can we determine the presence or absence of mental process. There are not isolated "atomic minds" or "atomic living systems" but only process of interaction. Minds are not "things,"

"monads," "pure cognizing agents," or anything else of a separate and unitary nature.

2. The interaction between parts of the mind is triggered by difference. That is to say, one part of mind responds to a *difference* between two other parts, not to one of the other parts alone. For example, our own eyes are constantly making short movements. If somehow they could be kept absolutely still while we were looking at an unmoving scene, we would quickly become temporarily blind. That is, our eyes are actually responding to difference caused by their rapid movements. A simple example: it appears that a frog literally does not see a fly until it flies.

The important point here is that differences, too, are not "things." Nor are differences localized. It is the difference between the blackness of this print and the whiteness of the page which provides meaning for us as we read, yet these differences are not themselves localized on the page. This differs radically from the idea of "cause" in physical systems, which can be attributed to the impact or force exerted by a part of the system on another part. Such a "cause" is always energetic—accompanied by the transfer of energy, and this is the only kind of cause recognized by physics. In the case of difference, there may be no energy exchange at all. Bateson points out that an event that does *not* happen, such as our forgetting to send in our tax forms, can make a difference (which can alert the auditors).

3. The energy for the mental process is triggered or fired or released by the energy of the change; the latter energy does not feed directly into the mental process. For example, the energy of the frog's jump comes from the internal processes of the frog and is triggered by the motion of the fly. The fly does not directly supply energy to the frog until the frog eats it. This supplies energy for later jumps, but it is at a lower level of order than the energy by which the frog jumped, until the frog has digested the fly and transformed the energy of nourishment into energy stored in the order of its neural and muscular system.

Thus the energetic relationship between parts of a mental system are interlocking relationships between self-actualizing subsystems which go through their own internally coherent changes of state in response to differences. There may be a wide range of internal states available as response to any trigger. This emphasizes the difference between mental systems and inert "things" which are not self-actualizing. Bateson points to the "pathetic fallacy" of confusing the two either by, for example, attributing mindlike qualities to subatomic particles, or by attributing thinglike qualities to humans, dogs, and other animals. Such confusions may seem to be only styles of speech, but as we have seen, styles of speech may hide much deeper confusions. The first confusion, attributing qualities of "charm," "strangeness," and so on to particles, may be relatively harmless. The second, however, attributing thinglike, machinelike qualities to minds, could be dangerous, as well as being, according to this discussion, simply wrong.

4. Mental processes require circular (or more complex) chains of determination. This is a very important point. Bateson gives the example of a man chopping a tree with an axe. The mental or living process involved must be thought of as including the entire circuit: man's brain, muscular system, the axe, the cut in the tree, man's visual system.

Such circuits provide the dynamic stability and self-correction of mental processes. For example, each successive stroke of the axe is corrected in response to the result of the previous stroke in a completed feedback circuit. We cannot isolate a partial arc of one of these circuits and attribute all of the mental process to that arc. Bateson speaks of the necessity of taking into account the complete circuit, particularly in relation to our idea of the "self."

"The total self-corrective unit which processes information, or, as I say, 'thinks' and 'acts' and 'decides,' is a *system* whose boundaries do not at all coincide with the boundaries either of the body or of what is popularly called the 'self' or 'consciousness'; and it is important to notice that there are *multiple* differences between the thinking system and the self as popularly conceived:

a. "The system is not a transcendent entity as the 'self' is commonly supposed to be.

b. "The ideas are immanent in a network of causal pathways along which transforms of differences are conducted. The 'ideas' of the system are in all cases at least binary in structure. They are not 'impulses' but 'information.'

c. "This network of pathways is not bounded with consciousness but extends to include the pathways of all unconscious mentation—autonomic and repressed, neural and hormonal.

d. "The network is not bounded by the skin but includes all external pathways along which information can travel. It also includes those effective differences which are immanent in the 'objects' of such information. It includes the pathways of sound and light along which travel transforms of differences originally immanent in things and other people—and expecially *in our own actions*."[3]

5. "The relation between the differences and the mental

processes they trigger should be regarded as the relation between a map and its territory. That is to say, an object, event, or difference in the world 'outside' an organism triggers changes within the organism which are not identical in type to the outside events, but are analogous to them as landscape is to a map. As we have seen, the work of Maturana and Varela suggests that the relation between possible states of the organism, considered as an autonomous system, and the triggers in the "outside world" is somewhat different from the one-to-one correspondence, the "representation" of a territory by its map. In fact, even what constitutes a "trigger" depends on the previous history of the interactions of the organism and on the immediate conditions of the present interaction.

      6. Such relationships between changes or differences in mental processes form hierarchical levels of order. Each level of order involves the recognition of the context of the lower order, the context of the context, and so on.

      Bateson elaborates this notion of hierarchies of order in mental processes by describing different types of learning process.[4] The first level, Learning 0, is that of the amoeba which responds automatically, according to a genetically fixed response, to the presence of food or an irritant—it simply moves toward or away from it. It has no choice, and the "learning" has already taken place at the species level. The next level, Learning I, is that of Pavlov's famous dog. A bell rings whenever the dog is presented with food. The dog salivates at the sight and smell of food according to the level of learning of the amoeba. However, after several such incidents, if the bell is rung without food, the dog salivates. The dog has learned to recognize the *context* in which food is presented, the context being a ringing bell.

      The next level, Learning II, which Bateson calls "learning to learn," is illustrated in a delightful story Bateson tells about a dolphin. A female dolphin in Hawaii was being used by a trainer to demonstrate simple Pavlovian learning. The trainer would watch the dolphin, and when he saw her do something he wanted her to repeat (say, a tail flip), he blew a whistle and then fed her. Within three trials the dolphin would do a tail flip whenever a whistle was blown. She would then be sent back to the holding tank to await the next demonstration. Of course, in order for the demonstration to be effective, the trainer would have to select a new trick each time, which the dolphin would dutifully learn to repeat. Between the fourteenth and fifteenth trial, the dolphin was observed to appear very excited in the holding tank, swimming to and fro as if agitated. When she returned to the exhibition tank for the fifteenth time, Bateson says, "she put on an elaborate performance that included eight conspicuous pieces of behaviour of which four were new and never before observed in this species of animal." Bateson suggests that the dolphin, between the fourteenth and fifteenth sessions, had realized the *context* in which all these trials had been taking place, namely, that the trainer wanted her to learn new tricks.

The set of possible Learning II type responses available to a person constitutes his or her "character": dependent, hostile, anxious, narcissistic, energetic, bold, humorous, fatalistic, etc. All of these describe the patterns of a person's habitual transactions with his environment. For example, a "fatalistic" man might be one whose patterns of interaction with his environment could have been acquired through a history of Pavlovian or Skinnerian experiences. It follows that the type of world a person experiences himself to be living in is also determined by the history of his Learning II type interactions. Bateson comments, "No man is 'resourceful,' 'dependent' or 'fatalistic' in a vacuum. His characteristic, whatever it be, is not his but is rather a characteristic of what goes on between him and something (or somebody) else." Thus a person's character, as well as his feeling of "reality," arise as a result of his history of Learning II experiences. A person brought up in a predominantly Pavlovian type of environment will tend to structure the contexts of his perceptions and actions to perpetuate a Pavlovian or fatalistic "reality."

The final level, Learning III, Bateson suggests is likely to be rare even in humans. Something of the sort may occur in psychotherapy, religious conversion, meditation practice, and whenever there is profound reorganization of character. Quoting William Blake's "Without contraries is no progression," Bateson suggests that the creature is driven to Level III by "contraries" generated at Level II, that it is the resolving of such contraries that will constitute positive reinforcement at Level III, and that such resolution can take many forms. We may be reminded here of the possible responses of Prigogine's self-organizing structures to sudden increases of energy input. Einstein and Bohr both provide examples of the resolution of "contraries" in their scientific work and in their private lives.

Bateson speaks eloquently of the characteristics of Level III learning in relation to self-identity: "If I stop at the level of Learning II, 'I' am an aggregate of those characteristics which I call 'my character.' 'I' am my habits of acting in context and shaping and perceiving the

contexts in which I act. Selfhood is a product or aggregate of Learning
II. To the degree that a man achieves Learning III, and learns to
perceive and act in terms of the context of contexts, his 'self' will take
on a sort of irrelevance. The concept of 'self' will no longer function as
a model argument in the punctuation of experience."[5]

These levels of ordering should not be mixed. It is such mixing, or
collapsing of various levels into each other, that can give rise to mental
illness. This is the basis of Bateson's double-bind view of schizophrenia, which, he suggests, is the result of a person's being
repeatedly exposed to contradiction between direct, Level I,
messages, and the contextual, Level II, messages. Such collapsing of
levels is also responsible for tremendous confusion about the nature of
learning and of mental processes altogether. If we connect criterion 6
with criterion 4, the full description of such a level of learning or
"minding" must include the entire chain of determination, e.g., an
organism *plus* its environment. This, then, is a model of "living" or
mental process in which we have to say that such process is not
localized "within" the body. In fact, Bateson contends that "the
phenomena which we call *thought, evolution, ecology, life, learning* and
the like occur only in systems that satisfy these (6) criteria." We see that
in looking for evidence of mind or mental process we must include the
whole perceptual circuit. Bateson provides us with a way of understanding that mind itself is embodied in the higher order patterns in
nature. Mind is not a "something" separate from nature. It is identical
at various levels of order with all of nature, not solely with individual
brains. It emerges as a characteristic of processes of nature at a certain
level of evolution. It is therefore futile to look for evidence of mental
process as located purely in the brain of an individual organism. We
must look for such evidence in the entire network of patterns of
interaction which that organism has with its environment, or which a
group or society of organisms has with its environment. As Bateson
says, "The individual mind is immanent but not only in the body. It is
immanent also in the pathways and messages outside the body; and
there is a larger Mind of which the individual mind is only a subsystem.
This larger Mind is comparable to God and is perhaps what some
people mean by 'God' but it is still immanent in the total interconnected social system and planetary ecology." Bateson is not talking
here of a theistic idea of "A God" or "A "Mind." Rather, he is saying
that when we realize that mental processes are patterns in appearances, rather than entities, then these patterns cannot be localized
only in the human body.

The Fallacy of the Localized Mind

This localization of mind is perhaps the main fallacy in the idea that mind will be understood only in the circuits of the brain, analogous to a computer. Such insistence on the body as the sole locus of mind is a subtle return of the soul theory because it insists on the centeredness of thought and perception. Of course, it is more nihilistic than eternalistic in that it is usually agreed that the body is impermanent. Nevertheless, the insistence on trying to interpret perception in terms of a central locus is as anthropocentric as was the earth-centered model of the world before Copernicus and Galileo. And there seems to be no good grounds for it other than the personal wish to be at the center of things. We might note that among Bateson's six criteria of mental process, there is not one about a center. We believe that awareness is in our body because our perceptions seem to terminate there and because if we move from room to room our perceptions change while we feel they are in the same body. Thus we give primacy to the clear and distinct highlights of our immediate perceptions. But what of the awareness that there is something at all? We may question the assumption that the totality of awareness is located in the body. But if awareness is not localized in the body, is it localized in space-time at all? What could it mean to say that awareness is not localized in space-time? We should perhaps distinguish between awareness which is not localized and self-consciousness which is. We will return to this later.

There are other possible models of mind which seem to accord with what we know just as well as does the computer model. We could model the brain more on the lines of a television, than a video player. Suppose an intelligent person completely unfamiliar with twentieth-century technology and knowing nothing about electromagnetic waves were to examine functioning TV and video machines side by side. He or she would think that in *both* cases the picture and sound were created entirely within the machine. If we then told him that the TV set but not the video machine was picking up invisible messages from space, he would think either that we were referring to one of his gods, or that we were crazy. In this model the brain is one part of many loops which *are* thought or perception, and include all of the environment.

In Bateson's example of a man cutting down a tree with an axe, the whole circuit of man, tree, and axe must be regarded as the basis of mental process in that act. The man himself is only part of it. Thus, rather than say that the man is perceiving the tree and his arm swinging the axe, we could describe this situation in a way which would be more in keeping with the actuality of it by saying that the whole circuit is self-perceiving. The locus of this self-perception is felt by the man himself to be "inside" him. In our own culture he takes this locus to be in his head; in other cultures it is taken to be the heart or the *hara,* the abdomen. Yet any act of perception always includes some aspect of the world. Without the world the man could not perceive at all.

Edwin Land has put this well: "Ordinarily when we talk about the human as the advanced product of evolution and the mind as being the *most* advanced product of evolution, there is an implication that we are advanced out of and away from the structure of the exterior world in which we have evolved . . . . This mechanism has no separate existence at all, being in a thousand ways united with and continuously interacting with the whole exterior domain. In fact, there is no *exterior* red object with a tremendous mind linked to it by only a ray of light. The red object is a composite product of matter and a mechanism evolved in permanent association with a most elaborate interlock—so that there is no tremor in what we call the 'outside world' that is not locked by a thousand chains and gossamers to inner structures that vibrate and move with it *and are part of it.*"[6] Land is, of course, no mere professional philosopher but an inventive genius who developed the Polaroid camera. He came to the conclusion we quoted as a result of a lifetime's work on perception of color and stereo vision. Thus, to think of mind as located in particular organisms is helpful for the survival of those organisms but false from the viewpoint of the whole of nature. As Bateson says, mind and nature are a necessary unity: "Break the pattern which connects and you lose all meaning."

We should emphasize that the patterns of organization which *are* mental process do not necessarily have to be patterns in "matter" as we think of it on the scale of ordinary human life. Such patterns could exist at scales as small as a single cell or perhaps even the genetic material. Such patterns could also exist on a galactic or cosmic scale. Physicist Gerald Feinberg and biochemist Robert Shapiro have argued that life could be based on interstellar plasmas, electromagnetic field energy, magnetic domains in neutron stars, and so on. We might note that the estimated number of stars in the Milky Way galaxy is almost identical to the estimated number of neurons in the human cortex. The possibilities for exchange of organization or information within and between galactic systems are great. The level of complexity and order in such systems could certainly be as great as that of the human brain. As Paul Davis, professor of physics at the University of Newcastle, says, "We could describe this state of affairs by saying that nature is a product of its own technology, and that the universe *is* a mind. Our own mind could then be viewed as localized 'islands' of consciousness in a sea of mind."[7]

There have been, throughout the ages, various views on the relation between mind and matter. Some have said that mind and matter are the same substance. This substance may have the qualities of matter, with mind being simply a complex arrangement of matter and subjective experience an illusion. This is the materialist view. Its most sophisticated expression today is the identity theory that mind is identified with certain functions of the brain and that subjective experience is simply the "inner" sense of this functioning which from an "outer" or "objective" viewpoint is seen to be merely the product of material brain activity. This approach is also very closely related to "realism," that is, the view that assumes the "reality" of an objective world entirely uninfluenced by the perceiving subject. Other theories, the idealist theories, say that the one substance has the quality of mind or ideas and that matter is merely an illusion, another idea. A third group of theories suggests that the one substance has the qualities of neither mind nor matter, but that these are accidental characteristics of a more fundamental substance. This is neutral monism proclaimed by Bertrand Russell in one of his phases, and by Spinoza.

An entirely different group of theories proposes that mind and matter are essentially separate, different substances. This is dualism, or pluralism, if we think there may be many mindlike substances. Descartes was the most influential proponent of dualism in modern times. He argued that mind substance and matter substance could not interact and that their only connection was through the beneficence of God. In more recent times, Karl Popper has argued for a pluralist view. He maintains that his World 2 (subjective feelings and ideas) and World 3 (objective theories, works of art, etc.) have existence independent from the world of physicality, World 1. With Eccles he tried to show how Worlds 2 and 3 might be connected to World 1 through the speech center of the brain.

Within the domain in which these philosophers choose to define the problem of "mind" and "matter," they resolve that problem more or less satisfactorily. Having conceptually divided the totality in a certain way between mind and matter, subjective and objective, they create further abstractions to heal the division. Such divisions have relative usefulness but ultimately no special validity. From the point of view I am presenting, all these theories make the mistake of looking for *substance*. Regarding "mind" or "matter" as some kinds of substances, they then try to decide what kind of substances these may be and how they may be related. I am pointing out that "mind," "matter," and "substance" are all patterns, levels of ordering that arise from unconditioned nature when a particular perspective is taken. We have met this view in Bohm's interpretation of quantum mechanics and will meet it again in Whitehead's process philosophy.

The networks or "patterns which connect" may *include* loops deep within the brain, but also go far beyond these loops. They extend to simple perceptual situations such as the man, his axe, and his tree. And they extend beyond these to networks involving one or more other individuals, as in conversing with a friend or in a therapeutic or teaching relationship. These patterns of connection, of mental process, also extend to interaction with intellectual structures, such as scientific theories, through language; to group situations, whether family, political, or professional; and to the entire mythocultural system of a society. For each of these situations our idea of what constitutes the "mind" of an individual must be entirely different, since it is defined by the different networks we, the definers, are focusing on. To illuminate the way in which "mind" can be variously defined by the circumstances, Victor Frankl, author of *Man's Search for Meaning*, uses the following metaphor: the shadow of a solid object, such as a cylinder, sphere, or cone can appear differently, depending on which direction we illuminate it from. (See Figure 11.) Similarly, the mind/nature duality appears in different ways depending on how the observer makes the division, which questions he is asking, which networks of interaction he focuses on.[8]

This, then, is a broad view of the partiality of the central conception of "I" and the false view it gives us of mind, nature, and perception because of this partiality. We are imprisoned within our brain, as Mountcastle says, only so long as we identify mind with self-consciousness localized in the brain. What we have uncovered up to this point from our consideration of experiments on the system of visual perception, in the light of reports from the meditative tradition, is that at a level of awareness which is more detailed in perception of time intervals than is usual, our experience would be discontinuous. The appearance of a continuous world "outside," a world of solid objects, is a result of conjectures which are averages of many moments of perception. This discontinuity, combined with the realization that

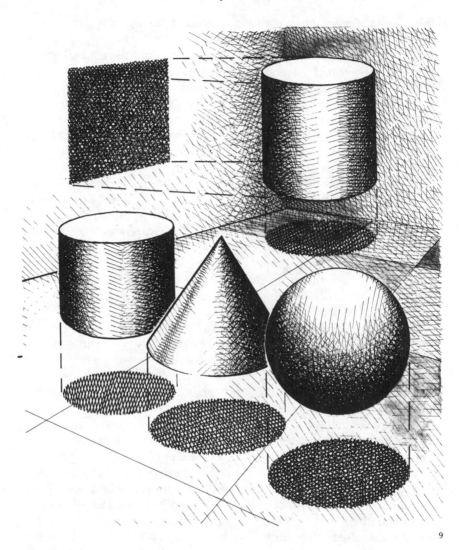

9

11.

the circuits of mental process extend beyond the organism, implies discontinuity of the supposed solid world as well.

I have proposed that the rapid flickering of our psychobiological states is a flickering back and forth between, at one moment, "actuality," form, specific conjecture about "the world," or the conjecture of "I" with a concurrent consciousness, and, at the next moment, that of which one cannot speak, *what is*, or a realm of unrealized potential. Specific moments in which we perceive our body and the material world alternate with moments of unactualized potentiality, and neither one is more primary.

We might note that this view of how experience of the world arises also cuts through the traditional philosophical categories of "realism" and "idealism." This view is not "idealism" because the unconditioned moments between each moment of form and awareness, and the creative process itself by which form and awareness arise, are not the products of an individual mind or stream of being. However, it is not "realism" because the forms that arise are conjectures arising within a particular stream of being, in response to unconditioned reality, out of the stored memory and experience of that individual stream.

This rapid flickering back and forth between our own self-organizing internal states and "that," whatever extends beyond those states and triggers them, is intimate experience which, because it happens so fast, is beyond the reach of observation by ordinary untrained attention. Thus, we have to use the personal language of contemplation if we are to speak about it directly, as it is experienced.

In a later chapter we will look at what has been discovered about the arising of thought and experience as a result of the direct observation of meditation practice. We will of course be using a somewhat different language: the language of personal experience rather than "objective" brain processes. But, nevertheless, we will be speaking of the same process of perception: the flickering between form and unconditioned; the way in which specific forms and their awareness arise from thought and memory; and the way the flickering is smoothed over by ordinary, coarse-scale self-consciousness to produce the appearance of continuity. In the next chapter I will outline this same process of perception as it has been described in the philosophy of Alfred North Whitehead.

14   Process and Feeling

Process philosophy was first proposed by that great English gentleman, mathematician and philosopher, Alfred North Whitehead, fifty years ago. Whitehead takes great care to show process philosophy to be a natural outcome of the Western tradition and to provide solutions to many of the seemingly intractable problems that had arisen in that tradition. It arises as a criticism of Berkeley, Hume, Locke, and Descartes, especially, but goes all the way back to find its roots in Plato and Aristotle. Whitehead himself is regarded by many as the greatest Western philosopher since Plato.[1]

Yet, in spite of Whitehead's demonstration that process philosophy is not antagonistic to the Western tradition, it nevertheless makes a profound break with this tradition. It is essentially a nondualistic philosophy in which the disastrous dualism between "mind" and "matter," "subject" and "object" is avoided at the foundation. There is therefore in process philosophy a very strong similarity to Buddhist doctrine, and it has in fact provided an important bridge between Buddhist and Christian as well as between scientific and Christian viewpoints.[2] We will particularly be discovering (in Chapter 15) a very close parallel between Whitehead's analysis of perception and that of the Buddhist Vajrayana tradition based on meditation experience. This parallel is all the more surprising since Whitehead does not speak explicitly of how he came across his understanding of process in perception, and he seems to have very little knowledge of the Buddhist tradition. Nevertheless, the quality and genuineness of his writing are such that one feels sure that he had direct experience of it rather than merely some theoretical acquaintance. Process philosophy is important for our discussion in that it provides the necessary link between the deeper understanding of perception that we are discussing and the abstraction and objectification of the scientific view. It is important also in that it demonstrates that the view of perception described in the Vajrayana tradition is not dependent on Buddhist doctrine as such. It is, rather, a cross-cultural discovery based purely on precise observation and analysis of the body/mind process.

Whitehead was trained as a mathematician and, at the age of fifty, he published with Bertrand Russell the revolutionary *Principia Mathematica*, an extensive and ground-breaking enquiry into the foundation of mathematics and logic. He then became dean of science at University College, London, and entered a very practical period of concern for education and educational administration, while continuing his work in mathematics. In 1924, at the age of sixty-three, he took up an appointment at Harvard University and lived in Cambridge, Massachusetts, until the end of his life. It was during this last period of his life that he developed the view of life known as "process philosophy." He appears to have been a very gentle and compassionate man as well as having extraordinary breadth of intellect.[3] This comes out very clearly in all his writings and his philosophy. Here is a description of him by Victor Lowe in *Understanding Whitehead:* "I cannot describe his face or recall any printed photograph or sketch that does justice to it. I can only confirm what Edmund Wilson wrote when he introduced Whitehead (as "Professor Grosebeake") into his early novel, *I Thought of Daisy:* that when you looked at him you felt that you were seeing a real face, in comparison with which others looked like mere masks. The general impression given by Whitehead's presence, I should say, was one of kindness, wisdom, and a perfectly disciplined vigor. Both his conversation and his writings showed a wonderful combination of urbanity and zest, rather like the tone of Plato's dialogues. He loved to follow the minds of young people, and when you came to him to talk about his philosophy, the meeting always began with the eager question, 'Tell me what you've been doing.' "

Process Philosophy

I will give a brief, summary overview of the main points of process philosophy which we will be interested in, and then we will go into each point more thoroughly. Whitehead regarded space, time, and matter as high-level abstractions from the immediacy of our felt experience. To mistake such abstractions as elements of concrete reality is to commit what Whitehead called "the fallacy of misplaced concreteness." Whitehead takes as primary the "unity of the perceptual field," which is "what it claims to be: the self-knowledge of our bodily events."

Such actual bodily events are moments of experience of finite duration, in which aspects of all immediately preceding events (the past) are "grasped into a unity." This grasping into a unity of aspects of the immediate past is at the level of "prehension" or feelings rather than conscious perception. The final occurrence of the unity of the felt moment is the emergence of value in the world.

Whitehead describes "all that is" as consisting of:

1. Actual entities, i.e., finite moments of experience.

2. Potentialities (which he also calls "eternal objects"), which constitute both what is and what is not but might have been. Each actual entity varies in the degree of inclusion in it of the potentialities. For example, a perception of a leaf might include "green" or "yellow."

3. The primordial nature of "God," which is an actual entity with a special status. The primordial nature is the unconditioned envisagement of the entire multiplicity of potentialities.

4. The consequent nature of "God," which is "the realization of the entire actual world in the unity of its nature."

5. A principle of creativity: each actual entity is self-created although it receives its lure and goal for self-creation, its aim toward satisfaction, from the primordial nature. Each actual entity aims for "satisfaction" or completeness, which is a relative grasping into a unity of feelings received from all the actual entities of its past. This is the process of creativity. When a given, actual entity achieves satisfaction as a "subject," it passes over to being "object" in the immediate past for the next actual entity.

Enduring objects, such as electrons or the individuality of persons, are due to repeating patterns of potentialities included in particular series of actual entities or "societies" of actual entities that prehend each other. If there is little or no novelty entering into such a series, it constitutes a physical object. The degree to which a series is "living" is the degree to which novel potentialities can enter.

I might add a brief note, at this point, concerning the apparent conflict between Whitehead's view that the "atoms" of reality are moments of experience and Bateson's view that there are no atomic minds and that to attribute "consciousness" to electrons is fallacious. Bateson appeared to believe that Whitehead was attributing "consciousness" to electrons, and that therefore they were in fundamental disagreement. Leaving aside the question of "consciousness," we seem to have two different levels of analysis here. Bateson is suggesting that at the level of ordinary perceiving and thinking, we make a mistake, and a very dangerous one, when we try to separate mind out, leaving nature as fundamentally mindless. Electrons are abstract entities derived by analysis of this "mindless" nature, and therefore must inevitably be regarded as themselves mindless. However, having recognized the necessary unity of mind and nature, and having recognized the arbitrariness of our distinction "outside world," we must then ask how we might analyze our experience without making this arbitrary split right at the beginning. If we agree with Bateson that we must extend our recognition of mental process beyond the boundaries of the body and of consciousness to include the "objects" and spatio-temporal relationships in the "outside world," then we might ask whether we were correct to separate out the objects and relationships as primary givens in the first thought. This is Whitehead's starting place.

Thus, Whitehead was able to show how, from a recognition of the primacy of felt, immediate experience, "things" and "living organisms" arise. He also showed, in great detail by a method he called *"extensive abstraction,"* how spatial dimensions could be derived. From this he went on to derive a theory of gravitation very close to that of Einstein but avoiding altogether the "fallacy of misplaced concreteness" with regard to space, time, and matter.

This, then, is a summary of some aspects of Whitehead's "process philosophy," and we will discuss each of these points again in more detail. It is extraordinarily difficult, in attempting to describe Whitehead's view, to gather his thoughts into a neat linear package of exposition. This was not the way Whitehead himself wrote, precisely because he was attempting to use language to describe a world which he recognized is essentially not linear and not at all similar to a simple subject-predicate construction. As Whitehead says, his theory implies "the abandonment of the subject-predicate form of thought, so far as concerns the presuppositions that this form is a direct embodiment of the most ultimate characterization of fact. The result is that the substance-quality concept is avoided; and that morphological description is replaced by descriptions of dynamic process."[4] Whitehead points out that the subject-predicate form of language (e.g., "The tree is green") is derived from our acquaintance with everyday things which, roughly, can be regarded as enduring objects with changing qualities. For example, in the summer we may say, "The tree is green," while in the autumn we may say, "The tree is golden," and in the winter, "The tree is bare." This gives us the feeling that there is a tree, the enduring object, with changing qualities such as the presence or absence of leaves and their color. However, as

Whitehead says, characterizing this traditional view, "Such an account of the ultimate atom, or of the ultimate monads, renders an interconnected world of real individuals unintelligible. The universe is shivered into a multitude of disconnected substantial things, each thing in its own way exemplifying its private bundle of abstract characters which have found a home in its own substantial individuality. But substantial thing cannot call unto substantial thing."[5] That is to say, if an object is essentially unchanging, it can have no real relationship with another such object. We have, of course, met with this issue before in our discussion of the role of language in forcing particular categories on our perception. We should remember that Whitehead was writing in the 1920s and 1930s, when language philosophy was in its infancy.

Whitehead's style of presenting his philosophy, reflecting his sense of process and interconnectedness, was more spiral or weblike than linear. Terms are introduced early and their meanings gradually built up through continual redefinition and cross-reference. Whitehead himself was extremely precise in his use of words, but in order to understand him precisely it is almost necessary to know the whole of what he is saying before one begins.

Whitehead wished to bring together within a single vision certain views of the world that had traditionally been regarded as somehow in opposition: permanence and change, continuity and discontinuity, many and one, matter and mind, subject and object, immanence and transcendence, the world and God. In particular, Whitehead criticized the "enduring object with changing qualities" mode of describing reality, the "elementary particle" mode, which he regarded as high abstraction from our immediate experience which is experience of *process* or change. And it is entirely fallacious to try to analyze change further into "elementary" objects that do not change. The tradition erred when it assumed that "process can be analyzed into composition of final realities, themselves devoid of process." The question for Whitehead then becomes not how to explain change in a world of unchanging things, but how to understand enduring patterns in a world of process.

There is another side to this question of change and permanence, which also comes from a fallacy derived from our uncritical observation of everyday life. This is the assumption that change is continuous. We see a bird fly, a person gradually become older, our thoughts flowing like a stream. We think of these as continuous changes and imagine that this is "ultimately" how things change. However, more refined observation of our immediate experience shows us that this experience is epochal or momentary. Change occurs as a moment or epoch becomes completed and passes into the next moment of experience in process of becoming.

Now, for Whitehead the world cannot be disjoint, requiring separate explanations for separate realms or aspects of existence. This view is shared, of course, with both materialists and idealists: the former try to explain the entire world including mentation on the basis of conformations of "matter"; the latter try to explain the whole world as mentation or ideas. Whitehead avoids both of these abstractions in asking what is the actual nature of human experience and proceeding from there. Thus, the fundamental fact of our human experience, that it consists of units or moments of becoming which pass into each other, must be the fundamental fact of all existence. I should emphasize that we are not necessarily talking about *conscious* experience, but rather experience at the level of feeling. We will return to this point later.

For Whitehead, then, all that is actual consists of epochal moments of experience, which he called actual entities, or groups of such entities. Each actual entity is itself a process of becoming; that is, there is a sequence in the process by which an actual entity comes to completion. Yet temporally each actual entity must be regarded as a whole—the process of becoming is not temporal. An actual entity cannot be temporally divided as it can in analysis. The world process consists of complete temporal units succeeding each other. Each actual entity arises out of an active process of self-creativity, which is common to all actual entities. Thus, this process of creativity, by which actual entities become complete, moment by moment, and pass into succeeding actual entities, could be regarded as the "ultimate" character of the universe. "Creativity" is the ultimate ground for the existence of the universe, which therefore does not need a transcendent "creator" for its complete explanation. Although we should be careful not to identify ideas that come from very different contexts, we can nevertheless already begin to see the extent to which Whitehead in the 1930s anticipated attitudes that are only now beginning to enter the sciences. For example, the self-organizing structures of Prigogine, and Feinberg and Shapiro's definition of life as the activity of a biosphere, both reflect ultimacy of "process" and "self-creativity" rather than "matter" and "creator."

Whitehead's task, then, is to describe the process of becoming of one actual entity, and the process by which an entity that has attained completion, and has become actual, passes into another actual entity in process. There is a rhythm here which is the fundamental rhythm of the universe, and in particular of human perception. It also anticipates the quantum nature of matter, the vibratory aspect of elementary particles.

"God"

Before analyzing the process of becoming, we must first consider an actual entity with a special status, which Whitehead calls "God." "God" is not the ultimate principle of the universe, which we have seen is self-creativity. Nor is it a transcendent creator, which we have seen is not needed. "God" is an actual entity, and in some sense is to be treated on the same level as all other actual entities. However, the special nature of the actual entity "God" is necessary in order to provide a complete explanation of process. Whitehead called this actual entity "God" in order to connect it in human experience with the sense of freshness and compassion which some people associate with religious or spiritual experience. And we will see how this connection can quite naturally be made. On the other hand, in conventional Judeo-Christian terms, the name "God" *is* associated primarily with a transcendent and all-powerful creator presiding over the world in judgment. Whitehead analyzed this conventional notion, as well as the opposing, pantheistic notion of a god that is merely nature itself, and showed that both are severely lacking. However, there are now deep-rooted misunderstandings embedded in our culture owing to popular associations of the term "God." It seems inappropriate to use it any further here, even in Whitehead's much more sophisticated sense. I will therefore refer to this particular actual entity by the two aspects, or natures, which Whitehead determines it to have: a primordial nature and a consequent nature. In brief—and we will return to this later—the primordial nature is the unconditioned envisagement, or realm, of the entire multiplicity of possibilities or potentialities which may enter into a moment of actuality; the consequent nature is the conscious realization of the entire, actual world as a unity.

The primordial nature will be more important to us for the particular aspect of Whitehead's view that we wish to describe, the process of perception. In order to understand this primordial nature, as well as to take another step in our analysis of the process of becoming of an actual entity, we need to take a look at the category of "potential." In any actual moment of experience, certain probabilities or potentialities which *could have* been included in that moment are excluded, while others are included. For example, "tree," "green," and "rustling" might be included, while "rock," "gray," and "thudding" might be excluded. In fact, we could perhaps say that the primary characteristics of a moment of experience are *which* particular potentialities are included in that moment and *how* they are included. By "how," we mean the valuation, intensity, or feeling quality with which a potentiality is included in an actual moment.

The primordial nature consists of a realm in which all potentialities, although not yet actual, are "envisaged" in their natural, hierarchically ordered relationships. Whitehead also refers to the "envisagement" of this realm as "conceptual valuation." We must be careful to understand that he is not using the term "conceptual" in quite the same way as we have been using it in this book, but more in the sense of "mental." If we regard the universe in its totality as having a "mental" pole and a "physical" pole, then the primordial nature would be the mental pole and the consequent nature would be the physical pole. We should emphasize that the primordial nature is not *before* all that is; it is *with* all that is as the primary fact which provides natural order and from which the self-causation of each individual, actual entity starts.

Experience

Each actual entity, a moment of experience, begins as a nascent subject aiming toward satisfaction, which is the self-enjoyment of its actuality as a completed moment. This subject arises, together with its subjective aim, from the primordial nature. The "objects" for this subject are the entire multiplicity of immediately preceding moments of experience. The term "object" should not be taken to mean an objective thing, eternally existing independently of the subject. Rather, "object" is always relative to the "experiencing subject." "Object" is not a synonym for an external thing but is always that preceding actual entity as it is *immanent* in the experiencing subject. In this way, Whitehead overcomes the absolute duality of subject and object that has plagued Western culture.

Precisely how preceding actual entities become immanent in a subject, and intensified and harmonized in that subject to form the final, complete moment of experience, *is* the process of becoming of that moment. Whitehead also refers to this process as one of "concrescence," or "growing together." The way in which each preceding moment enters into the concrescing actual entity is most analogous to feeling and to the direct transfer of energy that takes place in feeling. It is not analogous to a conscious apprehension, and for this reason Whitehead refers to this fundamental "entering in" of the object into the subject as "prehending."

To understand the importance of "prehending" as the mode by which the "subject" receives the "object," we must understand Whitehead's general analysis of human experience. For, as I have already pointed out, experiencing is the nature of all actual entities, and we can extrapolate from the nature of human experiencing to that of the "experiencing" of all actual entities. Whitehead criticizes very severely the view of experience that has taken root so profoundly in Western culture and which stems largely from Descartes. That is the notion that the predominant characteristic of experience is "mentation" or "consciousness," and that "unconscious experience" is a contradiction in terms. Of course, in the past fifty years with the advent of psychoanalysis and, following this, all kinds of psychological models, we are much more familiar with the idea of "unconscious experience." However, we still keep this understanding on the psychological level, while the sciences such as physics and biology are still built on the presupposition of the primacy of "conscious" experience—on the presupposition that "the clear and distinct sensation" should be the only basis of true perception and knowledge.

Whitehead points out that "thinking" and "consciousness" are variable—when we are asleep, knocked unconscious, or under the influence of drugs, there is no thought or consciousness. Therefore, as Leclerc points out, "Descarte's conclusion, 'cogito, ergo sum' is valid, but only as a conclusion from the fact of thinking to the fact of existing."[6] It is not, however, valid in suggesting that thinking is the essential *nature* of our existence or of our experience. Whitehead therefore rejects the sensationalist theory of perception, which identifies perception with *sense* perception. This theory holds that it is the consciousness of definite, clear-cut "sensa" which essentially constitutes perception. These "sensa" then are taken to be the fundamental elements in experience, all else being derivative from them. As Whitehead says in reference to this theory: beyond bare sense perception, given in immediate, present, and patterned connections, "the other factors in experience are therefore to be construed as derivative in the sense of owing their origin to these sensa. Emotions, aspirations, hopes, fear, love, hate, intention and recollections are merely concerned with sensa. Apart from sensa, they would be non-existent."[7] This view of perception is, of course, presupposed in the crude "camera theory" of perception, that the brain somehow constructs snapshots of an objective world that is immediately present.

Whitehead agrees that there are such sensa, but rejects the assumption that "because they are definite therefore they are fundamental." When we examine our experience, he says, "the first point to notice is that these distinct sensa are the most variable elements in our lives. We can shut our eyes or be permanently blind. None-the-less we are alive. We can be deaf. And yet we are alive. We can shift and transmute these details of experience almost at will.

"Further in the course of a day our experience varies with respect to its entertainment of sensa. We are wide awake, we doze, we meditate, we sleep. There is nothing basic in the clarity of our entertainment of sensa. Also, in the course of our lives, we start in the womb, in the cradle, and we gradually acquire the art of correlating our fundamental experience to the clarity of newly acquired sensa."[8]

This mode of perception, which is, then, not by any means the primary mode on which we must base our understanding of experience, Whitehead calls "presentational immediacy," because it is what presents itself immediately to our consciousness through our senses. However, he remarks, "clear, conscious discrimination is an accident of human existence. It makes us human. But it does not make us exist. It is of the essence of our humanity. But it is an accident of our existence."[9]

Whitehead then points to a more primitive, and more fundamental, mode of perception at the level of feeling, which he calls "causal efficacy." We can begin to understand causal efficacy when we acknowledge the mind-body unity and the "withness" of the body in all experience. We experience with the body. We do not merely see "red"; we see "red" *with* the eyes, we hear a sound *with* the ears, and so on. Whitehead points out that it is this "withness" that makes the body the starting point for our knowledge of our world. Therefore, the "withness of the body," rather than be dismissed as irrelevant, must form the foundation of our theory of experience as perception.

Yet we are not normally conscious of the bodily dimension, the "withness" of our body in experience. As Leclerc comments: "It requires a special direction of attention to become consciously aware of the bodily functioning in sensory perception. Yet it is there, for the clear conscious presentational perception is derived from the bodily sense organs. The essential point to notice is that 'derivation' is itself *unconscious* and does not enter as a conscious factor into the presentational immediacy."[10] It is, according to Whitehead, by understanding this unconscious role of the body, the withness of the body in perception, which is itself a mode of perception, that we begin to understand causal efficacy: "The causal influences from the body have lost the extreme vagueness of those which inflow from the external world. But, even for the body, causal efficacy is dogged with vagueness compared to presentational immediacy. These conclusions are confirmed if we descend to the scale of organic being. It does not seem to be the sense of causal awareness that the lower living things lack, so much as variety of sense-presentation and the vivid distinct-ness of presentational immediacy. But animals and even vegetables, in low forms of organism, exhibit modes of behaviour directed towards self-preservation. There is every indication of a vague feeling of causal relationship with the external world, of some intensity, vaguely defined as to quality, and with some vague definition as to locality. A jellyfish advances and withdraws, and in so doing exhibits some perception of causal relationship with the world beyond itself: a plant grows downwards towards the damp earth and upwards towards the light. There is thus some direct reason for attributing dim, slow feelings of causal nexus, although we have no reason for any ascription of presentational immediacy."[11]

Whitehead's viewpoint brings us to an understanding of the importance in perception of the vaguely felt but massive causal efficacy of the past at each moment. "Perception in this sense [causal efficacy] is perception of the settled world in the past as constituted by its feeling-tones, and as efficacious by reason of those feeling-tones." We usually describe our experience in terms of the immediate highlighting of particular sounds, smells, colors, and so on, or Whitehead's "presentational immediacy." But in doing this we ignore the connectedness of this present moment with every aspect of the immediate past, however dimly felt. For Whitehead, it was this ignoring of the embedding of our immediate impression in the vast web of causal links with the past that led to the imbalance in Western philosophy embodied in the distortion of scientific and analytic philosophy, with its emphasis on "clear and distinct ideas" as *the* criterion for truth. It is this embeddedness of a temporal moment in the totality of the actual world that brings to such a moment the sense of richness and value. "This mode produces precepta which are vague, not to be controlled, heavy with emotion: it produces the sense of derivation from an immediate past and of passage to an immediate future; a sense of emotional feeling, belonging to oneself in the past, passing into oneself in the present, and passing from oneself in the present towards oneself in the future; a sense of influx of influence from other vague presences in the past, localized and yet evading local definition, such influence modifying, enhancing, inhibiting, diverting, the stream of feeling which we are receiving, unifying, enjoying, and transmitting. This is our general sense of existence, as one item among others, in an efficacious actual world."[12]

Just as perception is the mode of causal efficacy, that is, perception at the level of feeling, is the primary constituent of our own experience, so "feelings" are the primary constituents of each actual entity or occasion of experience. "Feeling" is the basic nature of the act of prehending of one actual entity by another, by which the first becomes immanent in the second. "An actual entity is a process of 'feeling' the many data [past actual entities] so as to absorb them into the unity of one individual 'satisfaction.'"

Many Feelings Grow Together into a Unity of Satisfaction

I will describe now how the many feelings entering an actual moment of experience are unified in that moment. I will describe this process as simply as possible, while attempting not to distort Whitehead's view. This description may at first seem quite abstract, only because we are so used to thinking in terms of "things" in "space" and "time" and ignoring the presence of our body as the central reference of our experience. Ignoring the mind/body connection, we ignore our subtle feelings of the world.

The process of becoming, of concrescence or growing together of an actual entity, is a complex one that can be analyzed into a series of steps. Again we must remember that this is not a *temporal* series. The actual moment of experience is a temporal unit, of a certain finite duration. The series of steps is atemporal and, although to call it a series is accurate in analysis, in actuality all the steps take place simultaneously, within that duration.

First, there is the arising of a nascent subject as a subjective aim to satisfaction or completion derived from the primordial nature. With the arising of the subject, all the multiplicity of actual occasions in the antecedent world become "objectified" relative to this subject. They become data for the feelings of the subject. Second, the subject receives the feelings from its data, the antecedent occasions. The subject may receive feelings from the bare actuality of an antecedent occasion, pure physical feelings equivalent to the passing on of physical energy. Also the subject may receive feelings from the various potentialities that were actualized in the antecedent occasion. For example, suppose the antecedent occasion includes what we would in conventional parlance call a red ball. Then the subject will receive into itself a pure physical feeling of the sheer physical presence of the thing. It will also receive feelings derived from the qualities of redness, roundness, and rubberiness, all potentialities which had become actual in that particular occasion (whereas greenness, squareness, and hardness were potentialities which had not become actual in the antecedent occasion). A multiplicity of such feelings will be received by the subject from the multiplicity of potentialities actual in each one of its antecedent occasions. And a further multiplicity of such feelings will be received by the subject by virtue of the multiplicity of these actual occasions in its immediate past. Thus the nascent subject is receiving into itself a virtually infinite number of feelings derived from all the actual occasions in its past, and its task, its subjective aim, is to unify these into a final satisfaction. This process of unifying begins even as the feelings are being received: they are evaluated positively or negatively in relation to the subjective aim of the subject toward intensity and harmony.

The remaining stages in the process of concrescence of this particular moment of experience are all concerned with how the various feelings received from the "objective world," which have already been evaluated positively or negatively, are finally harmonized into the satisfaction of the subject. The third stage is a particularly important one. There are two possibilities at this stage: the first is that the feelings derived from the potentials in the antecedent occasions may simply recombine with the feelings derived from the bare physicality, without change. This corresponds to a simple repetition of the previous actual occasion, and is the type of activity of a simple "physical object" such as an electron. The other possibility is that new feelings may enter in, derived from the primordial nature, by comparison with the feelings from the antecedent potentials. This is the way in which novelty may enter into a series of actual occasions. The new feelings of potentialities derived from the primordial nature are then integrated with the physical feeling derived from the previous actual entity to give rise to propositions in which this actual entity becomes generalized into an "it." For example, in the case of the red ball, a new feeling of greenness may enter in. The feelings of "redness" and "greenness" when integrated with the physical feeling of a "something" (the ball), give rise to the propositions "It is red" and "It is green."

For Whitehead, the importance of propositions lies not in their "objective" truth or falsity but in their function as "lures for feeling." In the case of the ball, the entering in of the proposition "It is green," although it may not be true, may provide the inspiration to paint it green. Here is an example given by Donald Sherburne. "Many people in a town may be aware of the existence of an empty lot in the center of town, but only one enterprising businessman may positively prehend the proposition indicated by the words 'restaurant on that corner.' At the moment he first prehends the proposition it is false. But this is not the important fact about the proposition. As a lure for feeling the proposition may lead the businessman to buy the lot and build the restaurant."[13]

Propositions are predicates waiting for a subject, categories waiting for an actual occasion to fit them. This "fitting" comes about when the feeling derived from a proposition is contrasted with physical feeling from an actual entity or group of actual entities which are capable of forming the subject of the proposition (for example, the ball). The physical feeling from this actual entity is the objectified "fact," while the propositional feeling is the "might be." The contrast between "fact" and "might be," a tension held as a unity in the concrescing actual entity, has the form of consciousness.

Let us consider another case that might shed light on the formation of propositional feelings and their further integration with physical feelings in consciousness. Suppose, while you are reading, there is a dripping tap. At a level prior to consciousness you hear the sound, and possibilities as to its origin form as propositions: "It is a dripping tap," "It is someone knocking." At some point, you notice both the noise and the proposition and ask, "What is that noise, is it a dripping tap or someone knocking?" and you conclude, "Oh, it is a dripping tap." The final phase, "Oh, it is a dripping tap," in which all the various feelings (including other feelings involved in that moment; in this example, the sensation of sitting on a chair, etc.) are gathered into a complex unity, is the "satisfaction" of that moment.

Once a moment of experience has achieved satisfaction, by intensifying, contrasting, harmonizing, and integrating into a unity all its received and novel feelings, it ceases to be a "subject." It then becomes an "object" for subsequent actual occasions.

We should note that not all the stages we have described are *necessarily* involved in every concrescence. In particular, there may be no novelty entering in, in which case the actual entity is a simple repetition of a prior actual entity. There may also be no consciousness entering in at the later stage.

Whitehead's analysis of the process of the becoming of a moment of experience is, of course, far more detailed, rich, and subtle than this bare outline. However the reader has perhaps some idea of the way in which feelings are the basic constituents of an actual moment; of the way in which novelty or freshness might enter in, derived from the primordial nature; of the high level and variability of *conscious* feelings;

and of the final feeling of completeness and unity of that particular moment. In Chapter 15 we will discuss the Vajrayana Buddhist doctrine of the process of momentary perception, derived from observation in meditation practice. We will find a strikingly similar series of steps: the bare appearance of subject-object duality; the evaluation positively, negatively, or neutrally of the objects by the subject; a stage at which freshness, a glimpse of unconditioned possibility, may enter in; categorizing below the level of consciousness; and, finally, consciousness.

Societies

In order to elucidate the appearance of enduring patterns in a world of process, Whitehead speaks of a "society" of actual entities, by which he means a group, or nexus, of occasions which are ordered among themselves and which share a common characteristic by virtue of the inheritance of that characteristic from each other. Societies thus have the characteristic of endurance in time; they are constituted by a temporal series of occasions. Thus societies are the bearers of value—certain characteristics are valued by their endurance in a society. The characteristic of a society may range from the simple repetition of a physical object, such as an electron, to the enduring "self-consciousness" of a human personality. Thus, Whitehead was able to show how, from consideration of the primacy of moments of felt, immediate experience, things and living organisms could arise as the repetition of patterns of form in societies of actual occasions in which more or less novelty is able to enter in. The endurance of a society of actual entities implies considerable attainment of value intensity on the part of each actual entity which is a member. This is especially so in the case of highly ordered societies such as those involved in "personal order." Thus the society of which an actual entity is a member provides the supporting context for that actual entity to arise.

As Whitehead says, "Thus a society is, for each of its members, an environment with some element of order in it, persisting by reason of the genetic relation between its own members. Such an element of order is the order prevalent in the society." Furthermore, of course, any society does not and cannot exist in isolation. It can exist only within a context or background of societies of wider order which are supportive of its existence: "But there is no society in isolation. Every society must be considered with its background of a wider environment of actual entities, which also contribute their objectification to which the members of the society must conform. Thus the given contribution of the environment must at least be permissive of the self-sustenance of the society. Also in proportion to its importance, this background must contribute those general characters which the more special character of the society presupposes for its members. But this means that the environment, together with the society in question, must form a larger society in respect to some more general characteristics than those defining the society from which we started. Thus we arrive at the principle that every society requires a social background of which it is a part. In reference to any given society the world of actual entities is to be conceived as forming a background in layers of social order, the defining characteristics becoming wider and more general as we widen the background."[14]

This then is how Whitehead speaks of the interconnectedness of all actual entities, or moments of experience, each form a part of the background, at various levels of order, for all others. All of this is very reminiscent of Bohm's implicate and explicate orders in which, for example, an electron is to be considered as a series of electronic moments embedded or hidden in the implicate order. The appearance of the electron in the explicate order is, then, the sequential manifestations of the series of electronic events or moments arising out of the implicate order in a temporal series. Bohm, too, envisages levels of implicate ordering of wider and wider generality. Furthermore, for both Bohm and Whitehead the particular patterns, or laws of regularity, that we detect in the level of order of this particular cosmic epoch may well depend on levels of order of even greater generality, and these laws could therefore themselves be subject to change. As Whitehead says, "There is not any perfect attainment of an ideal order whereby the indefinite endurance of a society is secured . . . . The favourable background of a larger environment either itself decays or ceases to favour the persistence of the society after some stage of growth: the society then ceases to reproduce its members, and finally after a stage of decay passes out of existence. Thus a system of 'laws' determining reproduction in some portion of the universe gradually arises into dominance; it has its stage of endurance, and passes out of existence with the decay of the society from which it emanates."[15] This process refers to any society, from that of a fleeting elementary particle to that of our own cosmic epoch—"the widest society of actual entities whose immediate relevance to ourselves is traceable." This epoch "is characterized by electronic and protonic actual entities and by yet more ultimate actual entities which can dimly be discerned in the quanta of energy." But in this cosmic epoch too "there is disorder in that the laws are not perfectly obeyed, and that the reproduction (whereby each electron and proton endures with long life) is mingled with instances of failure. There is accordingly a gradual transition to new types of order, supervening upon a gradual rise into dominance on the part of the present natural laws."[16] This paragraph is characteristic of the extraordinary breadth of perspective that Whitehead brought to process philosophy.

The Fallacy of Misplaced Concreteness

In Whitehead's description of the world, built up from actual moments of experience, there is no ultimacy to the notions of "space" and "time." Indeed, Whitehead considered these to be high-level abstractions from immediate experience. He considered the tendency, both in unexamined everyday perception and in the Western philosophical and scientific tradition, to take space and time as the most real "givens," prior even to experience itself, as a dangerous fallacy, the "Fallacy of Misplaced Concreteness." However, in order to connect process philosophy with the categories of everyday experience and with scientific modes of description, Whitehead must show how space and time *can* be derived from societies of occasions. We have already seen how time may be understood as the ordering of a particular society of actual entities. "Time" is then relative to the particular society of entities we choose, and Whitehead was able to show that this definition of time was in accordance with the principles of relativity. He also developed a method of deriving spatial extension, which he called the method of extensive abstraction. This consisted essentially of defining related societies of actual entities according to a hierarchy of inclusion, rather as a set of concentric spheres or of Chinese nesting boxes forms a spatial hierarchy. With these methods of defining spatial extension and time, Whitehead was able to derive a theory of gravitation very similar to the General Theory of Relativity which nevertheless avoided altogether the fallacy of misplaced concreteness.[17]

I would like to mention here one recent proposal, albeit quite speculative, which attempts to extend the boundaries of science and for which process philosophy can provide a foundation and context. This is the theory of morphogenetic fields. This daring proposal has been put forward by Rupert Sheldrake in his book *A New Science of Life*.[18] He proposes the existence of a morphogenetic field of as yet unknown nature. This field is such that once any form has come into existence, it begins to make an imprint (a chreode) on the morphogenetic field, which will then facilitate the appearance of that form or one closely similar to it on a future occasion. It is rather like the channels caused by snow melting on a mountain; the next year, melting snow will find it easier to run down the channels already made from previous years. The existence of such fields would explain some phenomena that have been long standing mysteries; for example, the way embryos unfold in meticulously sequential patterns in space, with various changes occurring with extremely precise timing, is very difficult to understand in terms of the information in the DNA alone. And Sheldrake's theory would also provide an explanation for some of the evolutionary phenomena we have discussed, such as the appearance of parallel forms, the apparent foresight of preadaptation, and the apparent tendency of forms to experiment in certain directions, as in the case of antlers. Needless to say, Sheldrake's proposal is very controversial, having been condemned by the orthodox journal *Nature* and subsequently supported by the equally reputable but more flexible *New Scientist*. Part of the problem is that Sheldrake has formulated his proposal so far in such a way that the morphogenetic fields would need to transcend space and time and therefore be outside the realm of science as it is currently practiced. Also, the causation involved in the formation of chreodes and in their guidance of the appearance of new forms is supposed not to be an energetic one. Sheldrake is thus trying to draw biology into a much larger realm of discourse.

From the standpoint then of process philosophy, space, time, and energetic causation are high-level abstractions and by no means the ultimate factors which any explanation must presuppose. We have seen that every actual entity must take into account, in its concrescence, *all* antecedent actual entities no matter how remote. The intensity with which a remote actual entity becomes immanent in the concrescing entity may be substantially enhanced by the positive evaluation of that entity. Thus, actual entities in a society characterized by a particular form, say an embryo, may be enhanced by patterns imprinted in the larger society forming the relevant background by prior instances of that society, that developing embryo. Sheldrake's "morphogenetic fields" become the relevant order in the larger background societies and do not need at all to be considered transcendental.

Sheldrake's proposal does make experimentally testable predictions. The first data Sheldrake found in support of the presence of morphogenetic fields or patterns were in experiments done in 1920 by William McDougall. McDougall was testing learning in rats by running them through a maze and seeing how many trials it took them before they were able to make a perfect run. Twenty-two generations of rats later, he found that rats descended from parents selected for being *generally* slow learners were running this particular maze almost ten times faster than the first generation had. Sheldrake's explanation is that the learning of the earlier generation of rats was somehow imprinted in the morphogenetic fields and transferred to the later generation. When McDougall's experiments were later repeated in Scotland and Australia by rats *unrelated* to the original learners, they too mastered the same maze as fast as McDougall's originals. Sheldrake has suggested other experiments based on the morphogenetic field hypothesis, and these are now being tested.

Wholeness and Peace

We should now return again briefly to that special actual entity which Whitehead calls "God" but which we might perhaps better call the wholeness of the universe. For Whitehead, wholeness cannot be treated as an exception to all metaphysical principles, invoked to save their collapse. It is an actual entity which, although having a special character, should nevertheless follow the same principles of process as all actual entities. The primordial nature is one aspect of the special actual entity, but it is not itself an actual entity. That it is deficient can be seen in two ways. First, we have spoken of it as the "realm" of all potentials, yet the "realm" has to be "somewhere"; it cannot be separate from the universe itself. This "somewhere" is the universe considered in its unity; that is, it is the special actual entity. Second, in the primordial nature there is no becoming, no process, and no physical feeling derived from all other actual entities, all of which are necessary to an actual entity. The consequent nature prehends other actual entities in unity with the primordial nature. Thus, "the universe includes a threefold creative act: (i) the infinite conceptual realization (primordial nature), (ii) the multiple solidarity of free physical realizations in the temporal world (the totality of actual occasions), (iii) the ultimate unity of the multiplicity of actual fact with the primordial conceptual fact (the consequent nature.)"

While the primordial nature is eternal, unchanging, the consequent nature is temporal; it is "consequent upon the creative advance of the world," and it changes as it draws up into a new unity each new actual occasion in the world. Finally, because the consequent nature has physical feeling derived from all the separate actual occasions and infinite feelings of potential derived from the primordial nature, the consequent nature is also conscious: "The [consequent nature] originates with physical experience derived from the temporal world, and then acquires integration with the primordial side. It is determined, incomplete, consequent, 'everlasting,' fully actual and conscious." The consequent nature brings to the whole a sense of fundamental goodness, because through this consequent nature all individual actual entities are realized in a unity. Actual entities as subjects of individual experience arise and perish. But in their perishing they find their place in the integration of the ever-changing world. "In it there is no loss, no obstruction. The world is felt in a unison of immediacy. The property of combining creative advance with the retention of mutual immediacy is what is meant by the term 'everlasting.'"[19]

We might tentatively, at least at the theoretical level, correlate this special actual entity, wholeness, with the notion of unconditioned goodness which I introduced in Chapter 1. In particular, the unconditioned aspect could be partially correlated with the primordial nature as at once the ground from which a moment of perception arises and the source of novelty and freshness in perception. We should remember, though, that the unconditioned in its fullest sense is beyond the concepts of existence and nonexistence. It is not clear that process philosophy reaches this profound level of self-criticism.

While process philosophy provides a view in which the journey of human growth and spiritual training may be understood, Whitehead himself did not speak of such training, nor did he speak directly of the way in which his own insight and wisdom unfolded. This is no doubt in part due to his own humbleness, but it also reflects a shortcoming of the whole Western tradition of philosophy: a divorce of the theoretical view from the practicality of human transformation. Nevertheless, one has the impression that Whitehead understood completely the implication of a view such as that of process philosophy for the nourishment of humanity. His wisdom and compassion shine through all his writing, as in this passage on peace: "The Peace that is here meant is not the negative conception of anaesthesia. It is a positive feeling that crowns the 'life and motion' of the soul. It is hard to define and difficult to speak of. It is not a hope for the future nor is it an interest in present details. It is a broadening of feeling due to the emergence of some deep metaphysical insight, unverbalized and yet momentous in its coordination of values. Its first effect is the removal of the stress of acquisitive feeling arising from the soul's preoccupation with itself. Thus Peace carries with it a surpassing of personality. There is an inversion of relative values . . . . It is a sense that fineness of achievement is as it were a key unlocking treasures that the narrow nature of things would keep remote. There is thus involved a grasp of infinitude, an appeal beyond boundaries. Its emotional effect is the subsidence of turbulence which inhibits. More accurately it preserves the springs of energy and at the same time masters them for the avoidance of paralyzing distractions . . . .

"The experience of Peace is largely beyond the control of purpose. It comes as a gift. The deliberate aim at Peace very easily passes into its bastard substitute, Anaesthesia. In other words in place of a quality of 'life and motion,' there is substituted their destruction. Thus Peace is the removal of inhibition and not its introduction. It results in a wider sweep of conscious interest. It enlarges the field of attention. Thus Peace is self-control at its widest,—at the width where the 'self' has been lost and interest has been transferred to coordinations wider than personality. Here the real motive interests of the spirit are meant, and not the superficial play of discursive ideas. Peace is helped by such superficial width, and also promotes it. In fact it is largely for this reason that Peace is so essential for civilisation. It is the barrier against narrowness. One of its fruits is that passion whose existence Hume denied, the love of mankind as such."[20]

The extraordinary power of process philosophy is that it is able to provide a link from the most fleeting moment of immediate experience to our unexamined everyday world of things, as well as to the highest abstractions of science and the profound richness of human experience expressed in poetic and religious insight.

15 Vastness in the Minute Particulars

In this chapter, we will again describe the arising of a moment of experience, the arising of a moment of being from the ground which is beyond both being and nonbeing. This time, the description will be drawn from the meditative tradition of Buddhism.[1]

According to the meditative traditions of Shambhala and of Vajrayana Buddhism, manifestation and awareness of that manifestation are not separate, nor is one primary and the other secondary. They arise together out of the basic, unconditioned ground of goodness. The observation of states of mind and awareness in meditation practice—that is, bare, unconceptualized observation—indicates that manifestation and awareness arise together discontinuously, abruptly. Each individual moment in awareness is also an individual moment in manifestation, and between such moments there is nonmanifestation, nonexistence, a return to the unconditioned gound in which such exclusive categories do not apply. Each such moment is self-formed owing to prior patterns of conditioning. Such moments of manifestation and awareness are said, traditionally, to be in duration approximately one-sixtieth of the time it takes to snap the fingers.

The description of the arising of such a moment of experience from within awareness is sequential. However, this sequential nature is not interpreted with reference to ideas about time. Within the apparent duration of each moment, there is a sequential arising of specific patterns. Those elements of experience which were formed by habit in previous moments are gathered together to form the apparent unity—with discrimination of specific characteristics—of this moment.

Unlimited Gates of Perception

The description begins at the unconditioned level, which is felt as unbounded, complete, open, and full of potentiality. Within this openness, the potentiality of sense perception is also unlimited. These unlimited bases of perception are referred to in Sanskrit as AYATANAS. *Ayatana* literally means "gate of coming into existence," or as Edward Conze says, "perhaps 'source' would be a tolerable equivalent, since 'gate' has the connotation of 'cause' or 'means' [which is not intended]."[2] There are twelve such *ayatanas* arising in pairs: seeing (eye) and field of sight, hearing (ear) and field of hearing, and so on for the other three senses, giving five pairs. The sixth pair is cognizing (thinking mind) and field of thought. At this level, then, the potentiality for seeing, hearing, and so on, and thought is unlimited. There is no boundary to the possibility of seeing, hearing, thinking, and experiencing. There are entire worlds, which we have not imagined, which are available at this level as potentialities for perception. As Chögyam Trungpa says: "You experience a vast realm of perception [*ayatanas*] unfolding. There is unlimited sound, unlimited sight, unlimited taste, unlimited feeling, and so on. The realm of perception is limitless, so limitless that perception itself is primordial, unthinkable, beyond thought. There are so many perceptions they are beyond imagination. There are a vast number of sounds. There are sounds you have never heard. There are sights and colors that you have never seen. There are feelings that you have never experienced before. There are endless fields of perception.

"[But] because of the extraordinary vastness of perception, you have possibilities of communicating with the depth of the world . . . . the world of sight, the world of sound . . . the greater world."[3]

The beginning of a moment of experience is, then, that awareness notices itself floating in that vast space of possibilities and begins to perceive itself as an echo or reflection within that. Noticing such echoes, awareness realizes the limitlessness of the gates of perception. From each minute echo of itself, awareness recognizes the vastness within which it arises.

At some point, a sense of separate identity appears, and with this separateness fear arises. Why such a feeling of separateness arises cannot be said; it can only be seen. It seems to arise spontaneously without apparent cause, other than habit. Separateness once having arisen, the identity, instead of relaxing back into openness, separates itself further. Fear increases, the awareness begins to reduce itself, to limit itself, to try to ignore the fear and separateness.

Between the unlimited ground of awareness, called the *alaya*, and the unlimited sense perceptions, an intermediate level appears. This is known as the seventh consciousness (the first six corresponding to the six pairs of *ayatanas*). This seventh consciousness has an inquisitive, intentional, fearful quality. It acts as an intermediary between the *alaya* and the sense fields. It takes the bare sense perceptions and puts them together with its own version of *alaya* to form a story line, a sense of continuity, of personal history. The seventh consciousness is like an editor, which is continually taking in impressions, storing them, categorizing and editing them according to past habit and memory and playing them back.

Another analogy of the seventh consciousness is a fierce wind blowing over the surface of a lake (the *alaya*). The lake is stirred up so that it cannot clearly reflect the sky and moon, mountains and trees. In a similar way, seventh consciousness stirs up the basic goodness of *alaya* so that it cannot clearly reflect the vast space of *ayatanas*. As Chögyam Trungpa says: "In [examining] the process of perception, first you have a sense object. Then you have the actual mechanisms which perceive things, your physical faculties. You have eyes and ears and so forth. Beyond that is the mental faculty which uses those particular instruments to reflect on certain objects [this is the cognition, the sixth *ayatana* pair]. If you go beyond that, there is the intention of doing that, the fascination, the inquisitiveness that wants to know how to relate with those situations. And if you go back beyond that altogether, you find that there is some basic experience which is known as the alaya principle, which is basic goodness.

"The natural goodness of alaya, basic goodness, refers to experience instead of simply to the structural mechanical process of projection. We could describe that process with the analogy of a film projector: First we have the screen, the phenomenal world. Then we project ourselves onto that phenomenal world. We have the film, which is the fickleness of mind [seventh consciousness], which constantly changes frames, so we have a moving object projected onto the screen. There are a lot of mechanical devices to make sure the projection is continuous—which is precisely the same situation as the sense organs. We look and listen. When we look, we listen. We connect things together, although things are shifting completely, every moment, by means of time. Behind the whole thing is a bulb which projects everything onto the screen. This is like the nature of alaya, brilliant and shining. It does not give in to the fickleness of the rest of the machine."[4] The seventh consciousness, then, is restless, inquisitive, watchful. It is sometimes known as cloudy mind because of its nature of confusion and lack of clarity.

Rather than rest back into the *alaya*, the fearful mind of the seventh consciousness begins to narrow down to a limited world which is manageable. The perceiver begins to perceive particular limited perceptions, although such perceptions still carry with them the hint or flavor of the vastness. Because of this lingering hint of vastness, things have a symbolic quality; they are symbols of themselves, they bring with them their own sense of the unlimited goodness of *alaya*.

The *ayatanas* are said to be like gates that can open from both sides, that of the perceiver and also that of the perceived. Any particular object, in bringing with it the sense of the unlimited, can itself open further, reveal itself further. Whether it does so depends on the state of openness of the perceiver.

We are describing now the level of simple, direct perception of a particular sense field, perhaps a flash of a red window frame caught briefly as we travel past a village in a train, a sharp shadow of a leaf in the sunshine as we walk down the street, the sudden sound of a bell ringing, or the thick smell of cow dung, caught in the breeze. At this level there is no naming. There could be sights and sounds which we have never experienced before, for which we have no name or familiar description, and which may pass by unnoticed. Things are less frightening at this point and could be directly, truthfully perceived.

The Development of Duality

Next comes the first recognition of a world, an "other" which is separate and external to oneself. It is therefore also the first tentative recognition of one's "self," although it is still taking place far more rapidly than the level of what we usually call consciousness. In order to ignore further the fear of his separateness, or we could equally say, the challenge of this possible openness and vastness, the perceiver grasps tighter to his limited perceptions and begins to lose the freshness, brilliance, and gentleness of those perceptions that come from the reflection or aftertaste of the unlimited in them.

The development of this moment to its final stage as a conscious thought now continues almost automatically through four further stages. Because we have now made the first discrimination between "myself" and "the world," every further stage is an interpretation of the world in reference to myself. At this point, however, we do not yet have a clear sense of self-consciousness, simply a vaguely but definitely felt presence of "other" which almost indirectly implies "self." "Other" is now felt in its relationship to "self" as either supportive, threatening, or neutral, and there is an automatic reaction to these feelings as pleasurable, unpleasurable, or meaningless. Every aspect of our perceived world at that moment brings with it a certain quality of feeling, positive or negative, or it is simply not bothered with in that particular moment and fades into the dimly felt background, deliberately ignored because everything must have a place in relation to the nascent self. If we notice our own feeling as we look around a room, especially a room we are somewhat familiar with, we can catch a sidelong glimpse of this positive-negative tone that comes along with each perception. We have our favorite cup which always feels good, and the favorite chair we like to sit in, some memories that always come with a nice glow. Then we have certain rooms which always feel hostile, certain photographs which give us a bad feeling, and so on. All of this is continually taking place at a level before all self-consciousness or naming.

At the third stage, a definite central reference point is established. This is the point around which the entire moment is constructed. It is the first definite sense of "I," although it still has no name. It is a sense of criteria, judgment, measurement. This moment is beginning to take a particular shape in relation to the center; it is a particularly grand moment in which we feel very important or not important at all. In relation to "I," other begins to be felt as "outside," as near or far, as large or small. This is the point at which the sense of eternalism or nihilism enters in. This moment will confirm either the existence of "I" or the nonexistence of "I." Sometimes when we wish to say something, we find ourselves just saying, "I . . . I . . . er . . ." We seem to get stuck reflecting on our I-ness. This may be a hint of the third stage. It is also said that this is a point in the development of our experience at which we might be able to open again to unconditioned *alaya*. It is not necessary that our consciousness develop in a purely mechanical way, based on habitual patterns; it is possible that a sense of freshness and fundamental relaxation could enter again at this point. However, if this glimpse of openness does not come in, then the perceiver mechanically reacts to his feeling of "other" and its confirmation of him. Thus impulsive action based on habitual reaction happens.

Following the appearance of the central reference point, naming enters in. Perceptions are categorized and named according to all of the complex philosophical, psychological, and practical habitual thought patterns that are carried over from past experience. This is the level of language and concept. Still not yet conscious, it consists of an extensive web of associations and systems of thought: wholesome and unwholesome, religious and secular, all of our various opinions and prejudices that form the assumptions and preconceptions within which we fit our experience.

At the final, fifth stage, conscious thought arises, a conscious experiencing of something. It is this conscious level that continues in a constant stream of thoughts and conscious perceptions of all kinds. Because of this stream of thoughts which covers over and consolidates all of the previous stages, we do not normally notice these previous stages in our everyday experience. We do not see the arising and ending of each moment of experience; rather, our stream of conscious thoughts produces a sense of continuity to our experience. These thoughts may be thoughts about perception ("That is a red flower"); solemn, heavy, meaningful thoughts of a philosophical nature; light, flickering, unstable thoughts of a fleeting memory; or hunger pangs, and so on. But they are all tied together in such a way that our feeling of a solid, safe, reasonable world of commonsense objects, meaningful relationships, and all the rest is continually maintained. There are few gaps, few hints of freshness or openness.

This is the description of how each moment of ordinary experience arises from the unconditioned ground, according to the discoveries of meditation practitioners in the Buddhist traditions. Although it has been presented systematically in traditional texts, it is not essentially a philosophical doctrine but a formulation of direct observation. We have presented it informally in order to give the flavor of it. In traditional texts the five stages of development of experience are known as SKANDHAS, a Sanskrit word meaning "heaps" or accumulations.[5] This reflects the discovery that each stage arises simply as a heaping together of elements of habit passed on from prior moments. As we saw in the previous chapter, this description of the way a moment of experience arises to the level of consciousness is very similar to Whitehead's description of an actual occasion, a moment of experience.

Seeing Through Duality

We can see immediately that it provides the possibility for a different way of viewing and experiencing the ordinary world. We could begin to see through the stages of experience: past the conscious stream of thoughts and images, past the layers of conceptual systems, past the sense of central reference point and the feeling tone of perception, back to the bare sense fields. At that point we arrive back at the seventh consciousness itself, the cloudy mind which right from the beginning has been stirring up the habit patterns which produce the sense of continuous fearful and limited existence.

There are two different approaches that we can take to the energy of the *ayatanas* and the seventh consciousness, which are reflected in the differences between the Shambhala or the Vajrayana Buddhist traditions and religious approaches generally. The approach of religion is usually to boycott the seventh consciousness directly, to step beyond it and let our mind rest in the basic *alaya*. Cloudy mind is allowed to come to rest of its own accord. The whole process of formation of experience from its first separation from the *alaya* to conscious realization of thought and perception can go on, but now it is seen through. Experience can be seen as arising from *alaya* rather than separate from it. The gaps of nonexistence between experience are seen, and a sense of freshness can come in. Since from this point of view we are boycotting seventh consciousness, which ties our sense perceptions into a unified but separate world of experience, these perceptions themselves become suspect. A beautiful flower tends to be seen as a potential source of grasping and fixating onto the commonsense world, and we might try not to be caught up in it. The commonsense world is realized as insubstantial, empty, and we can see through it to *alaya*.

The Shambhala tradition, being a nonreligious tradition, a tradition of how to live in the ordinary world, takes a slightly different attitude to the *ayatanas* and the seventh consciousness. The sense perceptions are used directly and appreciated. Instead of being regarded as an obstacle to be boycotted, the seventh consciousness is purified and becomes a clear channel connecting the sense perceptions to the ground of basic goodness. Because the *ayatanas* were originally limitless in the vast ground of basic goodness, they are able to lead us back to limitlessness. We begin to realize the meaning beyond simple sense perceptions. The details of ordinary perception are bright, clear, and inherently pure.

From the Shambhala point of view, the key point is realization of the fear which narrows down the limitless *ayatanas* in the first place and perpetuates the mechanism. When fear is realized, then fearlessness can also be realized, and perception can be seen truthfully and directly. Thus the ordinary world regains its ground in basic goodness, and the natural wholesomeness of the world is discovered. In the Shambhala tradition this natural, inherent wholesomeness or healthiness of the ordinary world is termed sacredness. The secular world itself is sacred. Things do not have to be sanctified or made holy by something, someone, or some Great Being outside of themselves.

When they are seen fully and clearly in their own detail, then they are uncovered or rediscovered in their unconditioned dimension, basically not separate, good and, therefore, sacred. Such a sense of sacredness of ordinary, mundane experience transforms the world. Colors, sounds, smells, and so on are brilliant and clear, arising from nowhere and not dependent on anything, remaining just as they are.

This description of the process of perception is taken from the observations of generations of practitioners of mindfulness-awareness meditation who have been willing to look directly at their own minds and perception without the partiality of hope or fear. It shows how we form an apparently stable and continuous world from fleeting, unpredictable, and discontinuous flashes of perception. Notice that what we usually call "consciousness" comes in only at the final level of limitation. From this point of view, then, efforts to become "more conscious" might have the effect only of strengthening one's sense of limitation and separation, if these efforts are not based on insight into the nature of "consciousness" itself. This description begins to indicate that we might work back through the layers of narrowness and fear to rediscover the clarity and unlimited quality of the senses themselves and thus to begin to reconnect with the unconditioned goodness, *alaya*.

This working back through the process of perception and conceptualization, through the various stages by which perception concretizes the world, begins precisely where we are. That is, it begins by careful and detailed study of these processes at the ordinary, everyday level. However, such attention to, or mindfulness of, the details of the thought and perceptual process begins to lead automatically to a taming of the wildness and coarseness of the mind. Thus we begin to be able to be attentive to the smaller details of experience. This, then, is the process of mindfulness-awareness training, which is a natural and self-correcting process of joining mind and body.

Stages of Insight

According to Vajrayana Buddhism, as a practitioner accomplishes the process of mindfulness-awareness training, there are four main stages of perceptual discovery which he or she may glimpse. By stages of perceptual discovery, I do not mean that the student actually sees different things; rather, he sees things differently. His fundamental understanding, his way of seeing, and the nature of what he sees change. These stages are very clearly described in the Vajrayana tradition of Buddhism, from which the meditative training aspects of the teachings of Shambhala are mainly derived.[6] Some of the characteristics of these stages parallel the historical schools of Buddhism as they developed over the centuries: the Abhidharma schools, the Yogacara schools, the Madhyamika schools, and the Vajrayana or Tantric schools. Thus the particular stages of insight are to some extent represented by the formal doctrines of the historical schools.

The first stage is associated with the early pre-Mahayana schools and Abidharma doctrines, and particularly exemplified by the Sarvastivadins, a name meaning "those who believe all is real." Experience is recognized as transitory and discontinuous, consisting of a series of moments. Each moment is realized as analyzable into components, known as *dharmas,* or "knowables." It is the knowables that are real, and are direct objects of knowledge, immediately experienced. A moment of experience, which is ordinarily felt as a duality of a subject perceiving an objective field of objects in space and time, is seen to be composite. Each moment consists of a composite of *dharmas,* and the ever-changing flux of *dharmas* is what gives the appearance of change from moment to moment.

Each school during this period categorized these elements, knowables, *dharmas* and showed how any given experience could be analyzed in terms of them. It was almost like an atomic table of elements of experience. However, the specific number and description of the elements were not identical from school to school, although there was considerable overlap in detail and also general agreement as to the major subgroups of elements. The five main subgroups of knowables are elaborated by the Sarvastivadins thus:

1. The physical-material: solidity, cohesion, temperature, and movement, the five senses and the five sense objects and a sense of overall tightness or looseness, of cohesiveness or falling apart. This last element, cohesiveness or noncohesiveness, is what brings about the tendency of elements to stay together, or not, from moment to moment, i.e., the cause-and-effect relation between moments.

2. Mind, that which selects particular elements for attention. According to Herbert Guenther, "Perceiving and conceiving begin when we select and attend to connections which have already been 'thought' ['minded'] in sensory awareness, although they have not yet been 'thought about.'" This is the apparent subject of a moment of experience.

3. Mental events accompanying each moment of perception and coloring it. This is a category whose number of elements varies most from school to school. In one school, for example, there are

fifty-one[7] such elements, in another forty-six.[8] They include such knowables as attention, motivation, trust, nonhatred, lust, rage, opinionatedness, anger, jealousy, carelessness, drowsiness, regret, and discursiveness.

4. Connecting elements, claimed to be different from the "physical-material" and the "mental": nouns, verbs, sequence, time, process, distinctions, connection, birth, death, and so on. There are twenty-three such entities in one school, fourteen in another.

5. That which is not born from causes and conditions. This is the category of unconditioned *dharmas*. Three such *dharmas* were frequently described. The first is space, that which accommodates all that exists, yet is itself neither existent nor nonexistent. The second unconditioned *dharma* is nirvana, awakening attained through the practice of meditation, and the third is the natural, inherent, creative process which brings awakening spontaneously.

We might notice that at this first stage, the notion of an internal object, an "I," or of specific external objects such as "chair" or "cat," is not regarded as real and knowable. Such notions are hypotheses formed by mind (number 2) out of the physical-material elements and given their specific coloration and quality of existence by the mental and connecting elements. The experienced sense of selfhood in these schools was taken to be merely the aggregate of *dharmas* itself, changing from moment to moment. The possibility that freshness and openness might enter the stream of being is represented in this scheme by the fifth category.

There are obviously clear parallels here with the kind of work on cognition and perception that we have discussed, especially that of Gregory. The limitation that could come in at this point would be to take the *dharmas*, the knowables, themselves as ultimately real. These *dharmas* could be taken to have real, independent existence, permanent in past, present, and future. This early stage can also lead to a kind of personal nihilism in which wakefulness is regarded as individual annihilation. Or it could lead to a personal eternalism if the person were identified with category 2, the mind or subject of experience, and if this *dharma* were taken to be ultimately real and permanent.

The second stage of meditative insight into perception is to realize that all *dharmas* themselves arise out of a ground of consciousness, the *alaya*, and return to this ground. *Dharmas* then are seen to be not real, separately existing entities, but manifestations of the ground. Each experiential moment is itself taken to arise out of *alaya* as a result of the maturing of potentials deposited there from previous moments. The experienced sense of continuity, of ongoing life journey, is also this underlying consciousness, the *alaya*, which, however, has no individual sense of "self" associated with it. The conjecture of "I" comes from the stirring up of self-consciousness, which is the watcher and editor of experience.

Because all of phenomenal existence is seen to arise from and return to the *alaya* moment by moment, there is a sense of union or nonseparateness of phenomena. Thus, this stage is called the *yogacara* stage—the way of union (*yoga* = union, *cara* = way). Because the *alaya* does not have qualities of physicality or objectivity, there is, in this experience, a sense that the "relatively permanent conditions of interrelated senses are minds." Therefore this school, or stage of insight, is also sometimes referred to as the "mind-only" view. This term is very easy to misunderstand because, in the Western context, "mind" is always taken to refer to some individual, personal, localized mind. This is not the intention behind the phrase "Mind alone counts" or "mind only." The intention is rather to indicate that all of existence, all phenomenal appearance, appears at this stage to arise within and from a basic substratum of consciousness. Again, we are not referring to *self*-consciousness or consciousness of self. We are referring more to the creative process by which perceptions *become* conscious. Since at this stage phenomena are seen to arise from the nonseparateness of *alaya*, the acausal interconnectedness, or auspicious coincidence, of all phenomena is seen.

The problem which might arise at this stage would be an attachment to the idea of the *alaya* as ultimately real and having the nature of "self," and thus to "consciousness" or "oneness" as the absolute reality. The problem is that "All is One" can almost imperceptibly change to "All is Me" if one does not see the nonultimacy of both. There is still in this view a subtle watcher, an experiencer standing outside his experience and recording it as such. This error did appear historically as the "*citta-matra* error"—the belief that consciousness alone is real. Unfortunately, it is also beginning to be something of a problem in some current thinking, in which there is rather an overemphasis on "consciousness," especially individual consciousness. It is even sometimes proposed that "self-consciousness" is the next step in the evolution of man. This is the "*citta-matra* error" all over again. It creates the ultimate substantial reality of a Higher Self, or a Oneness, separate from the ordinary world.

Next, at the third stage, awareness begins to rest in its fundamentally unconditioned nature and to realize the whole process of conceptualization, including the conceptualizations of the "oneness" of subject-object, of mind-matter, of inside-outside, as just that: concepts. This stage, associated with *madhyamika* doctrine,[9] is often referred to as the realization of emptiness—a term which has been rather misunderstood by Western commentators. "Emptiness" is simply the fully experienced recognition of the entire structure of conceptualization and conjecture, centered on the central conjecture of "I." It is a recognition that *all* descriptions are relative and partial and that we usually live in our descriptions. Thus it is a realization of the fundamental, unconditioned, indescribable nature of reality. Instead of being felt as "oneness," it is felt as not-two. That is, having discovered the partiality and relativeness of the subject-object, mind-matter, inner-outer dualities, one rests in what is, neither one nor many, not-two. The discovery of emptiness is also an antidote to taking the dharmas as real, and this is how it was historically first understood.

However, there may still be some residual clinging to the "emptiness," to the experience of no-thing and no-self. Thus, in the fourth and final stage of Vajrayana doctrine, which is in some sense just the complement of the previous stage, the energy and brilliance of phenomena are seen. This stage unifies the fullness of the ground of consciousness alaya of the *yogacara* stage with the discovery of complete nonconceptuality, the "emptiness" of the *madhyamika* stage. A ground of awareness which is nonreferential is discovered beyond consciousness. Nonreferential awareness is realized as being unconditioned and at the same time the essential nature of man's being. It is this stage which prompted Guenther's remark, quoted in Chapter 2: "In order to discern the ultimate man must in some way partake of it." *Rigpa,* which we described in Chapter 2, is the faculty capable of discerning this unconditioned ground of awareness.

We have previously translated *rigpa* as intuitive insight. It might also be translated, according to Guenther, as "aesthetic experience" or "intrinsic perception," which terms emphasize its aspect beyond cognition. As Guenther says, "This [*rigpa*] is the aesthetic apprehension or intrinsic perception by the artist, the poet, the seer whose words are a commentary on a vision rather than a futile attempt to establish a system of universal truths over and beyond man's cognitive and sensible capacity, or to reduce the latter to some preconceived scheme demanding the exclusion of everything which the propounder of this scheme is unable to fathom."[10]

Guenther relates the content of *rigpa*, intrinsic nondual perception, to emptiness (nothing) and fullness of value: "In the aesthetic experience [of *rigpa*] we may say that that of which we are aesthetically aware, before we are aesthetically aware of it, is nothing. But this nothing, as the texts again and again assert, is not an absolute nothing which just is not. It is a dynamic nothing which, when we are aesthetically aware of it, has already been given a form and so resides in the vividly present and its meaning."

This stage, also known as "luminosity," is described by Trungpa thus: "If we see a red flower we not only see it in the absence of ego's complexity, in the absence of preconceived names and forms, but we also see the brilliance of that flower. If the filter of confusion between us and the flower is suddenly removed, automatically the air becomes quite clear and vision is very precise and vivid." He goes on to quote a Vajrayana text: "This energy, that which abides in the heart of all beings, self existing simplicity . . . is the sustainer of primordial intelligence which perceives the phenomenal world. It is indestructible in the sense of being continually ongoing. It is the driving force of emotion and thought in the confused state and of compassion and wisdom in the awakened state."[11] At this level, the simplicity and sacredness of the ordinary world are fully experienced.

Deepening Perceptions

We have gone through these stages as if each one followed only after the preceding one was fully accomplished, in a sequential fashion. However, experience *is* discontinuous, and the whole process of arising of conscious perception from the unconditioned ground occurs momentarily, over and over again. Therefore, it is quite possible for each of us to discover a glimpse of any of these stages at any moment. Such glimpses may be quite sudden and fleeting and easily ignored, but nevertheless they happen continually. The purpose of the practice of mindfulness-awareness meditation is to provide an environment which facilitates acknowledgment of such glimpses as part of our natural stream of being. It is because of such glimpses and the possibility of including them in our lives that our lives do not have to be automatic reaction, that perception can open up and humor and lightness *can* enter into them. To quote Chögyam Trungpa: "Your sense faculties give you access to possibilities of deeper perception. Beyond ordinary perception, there is super-sound, super-smell and super-feeling existing in your state of being. These can only be experienced by training yourself in the depth of meditation practice which clarifies any confusion or cloudiness and brings out the precision, sharpness and wisdom of perception—the nowness of your world."[12]

The change of perception we are talking about is itself a change of one's whole being. One begins to experience one's interconnectedness with the natural power and energy of the phenomenal world. Again to quote Chögyam Trungpa: "So meditation practice brings out the supernatural, if I may use the word. You do not see ghosts or become telepathic, but your perceptions become super-natural, simply super-natural.

"Normally we limit the meaning of perceptions. Food reminds us of eating; dirt reminds us to clean the house; snow reminds us that we have to clean off the car to get to work; a face reminds us of our love or hate. In other words we fit what we see into a comfortable or familiar scheme. We shut any vastness or possibilities of deeper perception out of our hearts by fixating on our own interpretation of phenomena. But it is possible to go beyond personal interpretation, to let vastness into our hearts through the medium of perception. We always have a choice: we can use perception to limit or close off vastness, or we can allow vastness to touch us. When we draw down the power and depth of vastness into a single perception, then we are discovering and invoking magic. By magic we do not mean unnatural power over the phenomenal world but rather the discovery of innate or primordial wisdom in the world as it is."[13]

Such transformed perception could be called "ordinary magic." It is ordinary because it is in this very ordinary, commonsense world in which we live, the world of cause and effect, of time and space, of life and death. It is ordinary because it is hidden from us by nothing other than our reluctance to see it, by habitual beliefs in struggle and separateness and by fear. It is magic because such transformation of perception happens suddenly, without cause or conditions even though the training may be long, requiring great practice and energy. Finally, it is magic because, when we discover nonseparateness from the phenomenal world, we also discover that perception and "innate primordial wisdom" are not two separate entities. The perceptions of this ordinary world, according to Chögyam Trungpa, contain wisdom within them. We will discuss this innate wisdom and power of perception in the final chapter.

16 Mindfulness and Awareness

The main theme which we have come back to throughout the book is that it is possible to work through the perceptual process and discover its nature and its foundation. It is possible in this way to bring about a fundamental transformation at the individual and social level that comes from recognizing unconditioned goodness. Many descriptions of this journey of discovery have been recorded both formally and personally in the neo-Confucian, Taoist, Buddhist, and Shambhala teachings. Also, one finds such descriptions, although rather rarely, in Christian-based contemplative literature. Meister Eckhart is the most clear example in the traditional literature, and Thomas Merton the most well known in recent times. According to Merton, "The first thing you have to do before you even start thinking about such a thing as contemplation, is to try to recover your basic natural unity, to reintegrate your compartmentalized being into a coordinated and simple whole and learn to live as a unified human being.

"The contemplative life is primarily a life of *unity*. A contemplative is one who has transcended division to reach a unity beyond division. It is true that he must begin by separating himself from the ordinary activities of men to some extent. He must recollect himself, turn within in order to find the inner center of spiritual activity which remains inaccessible as long as he is immersed in the exterior business of life. But, once he has found this center, it is very important that he realize what comes next.

"The true contemplative is not less interested than others in normal life, not less concerned with what goes on in the world, but *more* interested, more concerned.

"The 'reality' through which the contemplative 'penetrates' in order to reach a contact with what is 'ultimate' in it is actually his own being, his own life. The contemplative is not one who directs a magic spiritual intuition upon other objects, but one who, being perfectly unified in himself and recollected in the center of his humility, enters into contact with reality by an immediacy that forgets the division between subject and object."[1]

Perhaps one of the most extraordinary descriptions, in its clarity and matter-of-fact style, is the personal account of a fifty-year-old California housewife, Bernadette Roberts. After early experiences as a child, Roberts lived a Christian-oriented, Carmelite contemplative life for sixteen years and became convinced in her discovery of the fruition of that journey—oneness with God. She then married and had four children. However, twenty years later, a "second movement," as she calls it, began to happen to her in spite of herself. It was a journey to the complete loss of self and personal God as reference points or conceptual categories and the discovery of what *IS*. As she says, "Empirical reality is not itself an obstacle to seeing [what is], it is what we *think* about this reality that creates an obstacle to a transition that otherwise might not have been necessary in the first place." Further, "Self is but a temporary mechanism, useful for a particular way of knowing. Finding out what remains in the absence of self is the pearl of great price, a long journey, a change of consciousness and the beginning of a new life.

"By the time the journey is over, the only possible way of living is in the now-moment. There are no more head-trips—no clinging to a frame of reference, even if it is only the reference of tomorrow's expectation. What is to be done or thought is always under foot, with no need to step aside in order to find out what is to be thought, believed or enacted."

It is clear that for Roberts this journey was a spontaneously discovered one of natural awareness; for she says that, at a very early age, "by watching carefully I discovered that my feelings, emotions and certainly my thoughts, were quite separate and apart from something else that could leap and spread joy at some of the most inappropriate moments."[2]

My main purpose in presenting the concepts of science in this book has been to help to remove conceptual obstacles to recognition of unconditioned perception, obstacles coming from fixed limited beliefs about the nature of space, time, matter, mind, human nature, and the perceptual process itself. In addition, I suggested that what we now know about the physiology of perception is, at least at the theoretical level, compatible with the process and discoveries of meditation.

However, all of this would be rather beside the point if there were not a specific means, an ongoing discipline, by which each of us might make these discoveries for ourselves, personally. "Shall I compare thee to a summer's day, thou art more lovely and more temperate . . ."

makes little sense to someone who has not himself been in love or at least has a human body. A description of a rose is nothing if we cannot see and smell the rose and prick our finger on its thorns. Perhaps we could, just by chance, come upon these discoveries entirely unaided or just by thinking about them hard enough and believing in them. However, just as it is very unlikely that we will be able to sit down at the piano and play beautiful music without any training, so it is very unlikely that we will penetrate the process of perception without training. It is perhaps unfortunate that many of the recent books that argue so cogently for the need for a new vision place little emphasis on discipline, or do not mention it at all. It is a widespread misunderstanding that discovery of our basic nature is a "remote and fascinating experience."[3] Yet our original unconditioned nature is the closest to us of all that is. It is what we are when we relax our struggle for conscious control or for greater consciousness of any kind. It is not itself an experience at all, yet it underlies all our experience, even the most mundane. We can discover it in this mundane experience through practice of an awareness discipline. Practice simply clears away the obscuration which prevents us from resting in our unconditioned genuine nature and from perceiving and living in the world from that standpoint. It must also be clear that it is neither the intention nor the result of mindfulness-awareness practice to eliminate thought, as so many Western commentators have implied. On the contrary, as practitioners discover sooner or later, as one realizes the origin and nature of thought, so thought itself becomes clear and precise.

Shamatha: Mindfulness

As I have mentioned, the basic practice of Buddhist and Shambhala teachings is the practice of sitting meditation. This is also known by its Sanskrit term: *SHAMATHA—VIPASHYANA. Shamatha* means, literally, development of peace. It is also taming the mind or mindfulness. *Vipashyana* means insight or awareness. So this is also referred to as mindfulness-awareness discipline. In this chapter I will describe the development of mindfulness and awareness.[4]

Mindfulness is when mind is fully present with whatever action we are executing: placing a flower, wiping a teacup, washing the car, programing a computer. It is attention to detail, careful and almost deliberate. It is identifying fully with one's body, thoughts, and actions so that there is nothing left over, no self-consciousness, no watcher, no split mind. It is not watching what we are doing but simply *being* fully what we are doing, thinking, and feeling in its smallest, most insignificant detail. Awareness is the quality of sudden openness that comes in when we are fully present. It is a sudden glimpse, a sudden flash of freshness and wider perspective. We cannot discover where it comes from, we cannot hold onto it, and we cannot artificially recreate it. With this quality of openness comes a sense of inquisitiveness, of interest in the environment within which our actions and thoughts take place. Awareness might come in as a gap of openness in our solid train of thought or subconscious gossip. It might be suddenly glimpsing a flower or someone's face from a new perspective. It might be a touch of humor in the middle of a fit of anger. It is the spaciousness in our state of mind in which we realize that our thoughts, emotions, and perceptions are not solid, heavy "things," but simply transparent, energetic, and fundamentally wholesome.

In sitting practice, in order to give oneself the best opportunity for mindfulness and awareness to develop, physical activity is reduced to a minimum. In this way, perceptual stimuli become less crowded, and one has a chance to see the perceptual process in detail. That is, one simply sits, cross-legged, with erect posture. The back is upright and strong, the front open and soft, the head and shoulders strong, the breathing natural, eyes and all the senses open but relaxed, not focused.

As he or she sits, the student becomes mindful of posture and of physical sensation which may be restless, even occasionally painful, or quite relaxed. He becomes mindful of the physical sensation in his trunk, his limbs, and his skull, and of the slight movement of the chest and abdomen, and of the breath going in and out. The breath might be rapid or slow, shallow or deep, and we let it be as it is. Underlying this there is a fundamental quality of well-being and wholesomeness, the dignity of simply sitting there like a mountain without clouds. Body is simply there, solid, earthy, unmoving, and whole. Mindfulness of the solidity of body, of upright dignified posture, of the precision and naturalness of breathing forms the foundation for joining mind and body. This is known, traditionally, as the first foundation of mindfulness.

Next the student may become mindful of the feeling level of experience, which might be pleasurable, painful, or neutral. Feeling also might be particularly body oriented or oriented toward imagination and fantasy. Each physical sensation and each thought comes along colored by some mood, some quality of being pleasing or distasteful or merely boring. Going further into feeling, one might discover a basic sense of fear, a feeling of struggle for survival, of grasping for continued existence and fear that one may not survive. As one begins to let go of this struggle one discovers, with delight, that one *is* alive. It is a feeling of liveliness and perkiness which nevertheless one cannot cling to. One must simply touch it and let it go, not holding onto it but letting it be. When we realize this quality of liveliness, sitting practice becomes very personal and intimate. It is no longer a foreign idea that we have imposed on ourselves but a very immediate, natural expression of life as such. Although the practice may continue to be often boring and difficult, nevertheless, it has become a part of our stream of existence, and when we occasionally touch the sense of life it can be tremendously refreshing and a basis of natural confidence and delight. This is the second foundation of mindfulness.

A further aspect of mindfulness is the state of mind. One may find that one's mind is very restricted or quite expansive, very tight or rather loose. The mind might be in a state of elation or one of depression, concentrated or scattered. It might seem to be in a rather fine, almost spiritual state, or in a state of coarseness and turmoil. All of these are objects of mindfulness and are not regarded as particularly important in themselves. One might begin to notice also the apparent fickleness of mind, the flickering unsteady quality of thoughts and perceptions. Sometimes the student is able to be there, practicing the discipline, keeping the posture and mindful of body, feelings, and so on. At other times he might drift off, having no idea where he is, then suddenly return to mindfulness. One cannot find out what brought one back, but without any deliberate effort one realizes that one is sitting and returns to mindfulness. Acknowledging this sudden change of mind is the basis of effort in sitting practice. One cannot bring oneself deliberately back when one has drifted off, one can simply recognize that mind comes back naturally and abruptly. The student begins to develop a sense of trust that mindfulness is entirely workable, that it happens, in a sense, without him. One does not have to try to be mindful; rather, mindfulness enters one's being automatically if one sets up the general environment for it by sitting, with the intention to practice, and the appreciation of the possibility of practice. That mindfulness happens without our conscious or unconscious manipulation is a manifestation of our unconditioned goodness and is the reason that genuine mindfulness is possible at all. This is the third foundation of mindfulness.

Finally, the student might become mindful of the contents of mind, of all the variety of different types of thoughts, emotions, and sensations that are constantly streaming through our consciousness. There are quick, darting thoughts; heavy, somber thoughts; religious and philosophical thoughts; mean thoughts; angry, passionate, dreamy thoughts, and so on. He or she might also become mindful of the overall speed of the thought process: in a traditional analogy, sometimes it is dashing along like a rocky highland stream; at other times it is moving fast but contained like a river running through a steep canyon; still other times it may be moving very slowly like a broad river; and, occasionally, it might be calm and still like a lake without waves. One can be mindful of each of these qualities of mental processes equally. None is regarded as more desirable than another. They simply are the contents of one's mind as it is. The student discovers a further sense of his human dignity and confidence; he can sit there fully mindful, unmoving, and unperturbed as his thoughts go through constant changes. This is the fourth foundation of mindfulness.

The four categories of objects of mindfulness that we have just described—mindfulness of bodily sensation, of feeling, of state of mind, and of mental contents comprise all of our psychophysical being. Traditionally they are known as the four foundations of mindfulness.[5]

When one begins to practice mindfulness, one may find that things do not go smoothly. One may be very bored for a while, then quite restless. One may become quite angry and upset or quite depressed. Occasionally one may have a glimpse of fundamental natural dignity and confidence as one simply sits there, alert and relaxed. Then this glimpse seems to be almost immediately clouded over by more upheavals. Sometimes a glimpse of basic goodness may lead only to further despair, annoyance, or frustration at not being able to hold onto it or to relax into it and remain there. In order to help the student continue, traditionally, general types of obstacles or disruptive forces are pointed out with corresponding antidotes with which he or she might overcome these obstacles and continue.[6] The obstacles fundamentally have to do with the seeming heaviness and numbness of the body and mind and the tendency of discursive mind to try to avoid the insight of wakefulness, to want to cling to and perpetuate safe and comfortable habitual patterns of thought. Thus some obstacles have to do with laziness, forgetfulness, and general drowsiness and depression. Other obstacles have to do with wildness, with craziness of discursive chatter and imagination, with carelessness and general hypersensitivity, so that one is knocked off balance of mindfulness by the smallest thought. We might decide that to train our mind and discover our basic nature is an excellent idea, but we find when we actually begin to do them that we are beginning to take a look at our most personal, intimate experience, which we may wish to avoid.

Thus habitual patterns of body and mind begin to object to the intention to meditate and throw up obstacles. The way to work with these obstacles, the antidotes to them, begins with a general sense of trust in the efficacy of the practice and in the insight and dedication of millions of practitioners who have already accomplished the training. Beyond this there is the recognition of the need for effort in the training and for the development of a sense of familiarity with the training. If you wish to train yourself physically, by jogging or working out, you know that for the first few months it may be a diffcult grind. Later you find, to your surprise, that you are already beginning to become fit and in condition. At this point you realize that you are already addicted to the workout. It has become so familiar to you that you cannot let more than a few days pass without working out. It is precisely the same with mindfulness training. The discovery that the training is working, and the sense of familiarity with, and almost addiction to the training, can be recognized as antidotes to the laziness and depression that frequently come along. Finally a general, environmental sense of watchfulness can be an antidote to the wildness and carelessness of discursive mind, so that the mind's tendency to wander off can itself be a reminder to return to mindfulness.

Mindfulness practice, then, is the process of developing basic familiarity with the entire thought and perceptual process and a sense of friendliness to it. Mind is no longer felt as a hostile, strange thing, often distracted, in which lurk unknown terrors. We begin to tame and pacify the wild, restless energy of mind so that mind naturally begins to settle. The idea of "developing peace" is not a stopping of thoughts, but a discovery of the underlying peacefulness and breadth in which perception and thoughts arise, abide, and dissolve.

Traditionally, this gradual discovery of and resting in the fundamental peacefulness of mind is described in nine major stages, known as the nine ways of resting mind.[7] These stages range all the way from just beginning to draw in the wildness and uncontrolled thought process and prolonging that state a little; to being able to ride the thought process with mindfulness and without being swept off balance; to thoroughly resting the mind in peacefulness, as naturally as swans swim and birds fly. As one continues with mindfulness practice one develops a familiarity with one's thoughts and emotions. The doubts and fears, depression and wildness are less threatening. One begins to relax and feel fundamentally friendly toward oneself just as one is, without pretense. Thus one begins to reconnect with the softness and genuine tenderness of one's heart. Habitual patterns of thought and emotion which have formed a defense, covering natural tenderness, have provided a hard front to deal with the apparent harshness and difficulties of the world. But the rediscovery of tenderness begins to open one to the world and to the possibilities of clear perception and genuine communication.

In an analogy, the mind could be, at first, like a wild horse galloping in the meadows, running hither and thither, turbulent, energetic, and unridable. If we were to try to ride the horse of mind, it would throw us off or carry us away. First, we must tame it: we give it a very large pasture, and place the halter of technique around it with a very long rope. Gently we pull on the rope, just to hint that someone is there who would like to train it. We do not try to force ourselves into mindfulness; we simply give mind space. When the horse is close enough, we make a gesture of friendship, place a saddle on it, and mount. We can now guide the horse to manifest the dignity and elegance of dressage or to explore difficult mountain passes. Now that mind has settled a little and we have made friends with it, rather than try to hold onto the sense of calm, we can acknowledge the openness and liveliness that comes in. This is the *vipashyana* attitude.

Vipashyana Awareness

The awareness, or *vipashyana*, aspect of practice brings with it a sense of inquisitiveness, curiosity, liveliness, and insight. A traditional analogy is that while the peacefulness of *shamatha* is like a lake without waves or undercurrents, it might become stagnant if nothing grows in it. *Vipashyana* is like the lilies growing out of the lake and swans swimming on the lake.

Vipashyana develops as we begin to clearly discriminate the various perceptions, feelings, emotions, and thoughts that continually follow one another and intermingle in our lives. Once we have the clarity and brightness of awareness to begin to discriminate these, then a natural inquisitiveness is discovered, a natural interest in how they arise and enter into our experience. Because of this natural interest, and the steadiness of mindfulness that has developed, awareness is not disturbed by the large upheavals of emotions and moods that might still tend to take place and that are constantly coloring the state of mind. This is an important point in the development of awareness because it is precisely these emotional colorations that normally distract awareness and prevent us from recognizing the conceptualization we are projecting onto our experience. Out of this steadiness of not reacting to the large-scale ups and downs, together with precision of mindfulness, awareness of the minute details of the thought and perceptual process begins to develop. One begins to notice the smallest details of perception and thought, the relationship between them and the mood or mental atmosphere around each thought. One might also begin to notice the disjointedness or discontinuity of perceptions and the fleeting gaps between them.

In sitting practice and in its application in everyday life, the *shamatha* and *vipashyana* aspects are always present together. It is not a question of having to go all the way to complete *shamatha*, completely taming the mind, before any glimpse of *vipashyana* can come. Nevertheless, we usually find when we begin to sit that our experience of our minds is at first rather wild. Our minds wander between elation and depression, between uptightness and looseness, very rarely synchronized with body, which is just sitting there. Therefore, on the whole, *shamatha* practice is emphasized and *vipashyana* is allowed to develop naturally.

There are said to be six discoveries associated with developing *vipashyana*. These discoveries refer to the insight of direct perception, immediate experience. They refer, of course, not merely to new information about things, but to personal understanding. The first, the "discovery of meaning" is the discovery of how language works. The practitioner begins to discriminate the accurate use of words and logic, so as not to be confused by them but to get directly to the meaning conveyed by them. He sees the relationship of language to what is beyond language and is not confused by his own habitual thought patterns or those of others. He begins to be able to perceive what is going on beyond the verbalizations.

The second, "discovery of reality," refers to beginning to distinguish between "inside" and "outside." The practitioner distinguishes between personal opinion, emotions, or life situations generally and the larger world around him, from the weather to the state of international politics. He or she begins to distinguish what is a product of his or her own "inner" world and what is not. He also begins to see the importance of going out to others and of connecting with the goodness and wholesomeness of the world.

The third, the "discovery of nature," is beginning to see how thoughts rapidly follow the bare perception of what is, a color, a sound, and so on. The practitioner discovers how, from a first glimpse, thought of action or reactions follows. "Nature" here refers to the nature of the perceptual process, and also to the intrinsic characteristics or marks of perceptions which distinguish one from another.

The fourth discovery, "discovery of sides," is beginning to discriminate situations and actions that further mindfulness and awareness from those that do not. "Sides" here, then, refers to the rather commonsense meaning of "good" and "bad." This discovery is realizing what is helpful to oneself and others and what is not, almost at the level of ordinary decency and good manners. It is also being able to recognize when there is awareness and when there is not.

The fifth, "discovery of time," is becoming less confused by past memories and future hopes. The practitioner is able to discriminate past and future as they enter into the present moment so as not to be caught in the complicated pattern of hope and fear by which we spin our web of habitual patterns that color our perception. He knows, then, what he is actually experiencing now.

The sixth, "discovery of insight," is realizing and trusting in the causal efficacy and interconnectedness of the world. It is direct, penetrating insight into causal relationships of the relative world so that one begins to see how much is presupposed and taken for granted in our usual experience of the ordinary world. With this discovery the practitioner need not cling either to particular reference points for perception, or to logic in an attempt to sustain only his own viewpoint. He can begin to adopt various viewpoints without partiality. This discovery is the forerunner of *rigpa*, the intrinsic, nondualistic perception of unconditioned nature.

There is a great deal more detail concerning the journey of discovery of *shamatha/vipashyana* as it has been mapped out by generations of practitioners. I have tried to give the reader some flavor of the journey, enough to show that this practice is not some vague subjective introspection, but is a detailed method of training and evaluation which is intersubjectively accessible.

Mindfulness and Awareness in Action

It is important to emphasize that mindfulness-awareness practice is just that—practice. Practice is simply an opportunity to discover one's basic nature and to develop the gentleness and fearlessness arising from it, on which one's action in the world is based.

We have presented the principles of mindfulness and awareness in relation to sitting practice, in which physical activity is simplified to a minimum. In order to be active in the world we have to stand up, walk, and perform physical gestures and speak. There are other disciplines which form a very valuable link between sitting practice and ordinary action in everyday life. These disciplines, which come mainly from the Japanese traditions, might include *ikebana* (flower arranging), *kyudo* (archery), *chanoyu* (tea ceremony), calligraphy, various martial arts (such as *t'ai chi chuan*), as well as the Western descipline of dressage (horsemanship). Because they are themselves disciplines that have for centuries been ways to train the mind, they already embody the principles of mindfulness and awareness. Therefore, by practicing them in conjunction with basic sitting practice, we can begin to see how mindfulness and awareness can be carried on in physical activity, and we can then continue this attitude in our everyday life.

To bring out the natural living quality of harmony and elegance which comes from joining mind and body in mindfulness-awareness practice is the essence of the discipline of ikebana. Ikebana is a centuries-old discipline of arranging flowers which is at the same time both an expression of natural elegance and beauty, and a practice, a path (Japanese *do*), of training the mind-heart. In the more traditional arrangements there are three main branches representing heaven, earth, and man, and flowers representing the universal monarch. Writing of the traditional attitude to ikebana and its role in Japanese life, Gustie Herrigel says: "Now is such a flowerpiece a product of nature or of art? Or does it stand midway between the two, so that it is more than nature and not yet pure art? An unequivocal answer is extraordinarily difficult to give. For the Japanese, life and art, nature and spirit form an indissoluble unity, an unbroken whole. He experiences nature as having a soul, and spirit as part of nature, without purpose. So he cannot make sense of a question which presupposes a division of nature from spirit, life from art, as though they were alien to each other. For him nature is neither dead nor unspiritual, nor yet a mere symbol and semblance. The Eternal itself is immediately present in its living beauty. This viewpoint is typical of all Japanese art. Consequently, we fail to touch its real essence if we believe that it 'idealizes' its objects and aims at easing tensions and reconciling opposites in order to create harmony. For the Japanese, harmony is the innermost form underlying nature, life and the world, and art can have no other task than to portray this harmony, to confirm it through varying degrees of 'unconscious awareness.' The artist will draw it into himself as if with a deep breath from an infinite distance, exalt it, and body it forth. With his senses wide open, he perceives the new creation and carries it out of its background into visible form. Since he gives up all thought of placing himself in the foreground, he will, simultaneously with the tangible presence of the flower—in which the cosmos manifests itself—also become aware of the law of its being, and of his own nature."[8]

By following basic rules and arranging flowers and branches with attention to detail, the student trains his or her perceptions. The completed arrangement is not considered to be just an attractive decoration. It is a reflection of the student's state of mind, of how tight or relaxed, how scattered or attentive, how embarrassed or confident, how confused or balanced, he or she felt as each particular flower was placed. To take part in an ikebana class is almost magical. The beginning students make their arrangements, perhaps attempting to emulate an arrangement of the teacher. There is often an awkwardness in the arrangements. Perhaps they seem slightly cramped or overfull. Yet at the same time there may be a touch of delight, a hint of openness. A master teacher may then look at a student's arrangement, and with only a small change—a slight pruning here, a small turn of a branch there—suddenly the arrangement seems to come to life. What was, in the student's arrangement, simply a hint becomes a glowing manifestation of harmony and life. By practicing again and again to bring that quality of harmony and delight into his arrangement, a student begins to experience it in his own state of mind as well. Chögyam Trungpa, in speaking to the Kalapa Ikebana school, of which he is founder, had this to say: "This evening we are discussing perception and the appreciation of reality. Generally, we believe that it requires talent to create a work of art. People sometimes reject themselves because they feel they don't have such talent. Their art might be sewing, cooking, painting, interior decorating, photography, or anything that involves aesthetics. Of course, flower arranging is included. However, in this discussion, one's artistic talent is not the point. Anyone who possesses the appreciation of sight, smell, sound, and feelings is capable of communicating with the rest of the world. One's perception of the world, as well as a general sense of space, can be expressed in art. From this point of view, we could say that ikebana is a way to enter the general social world of our sense perceptions, and it is a way to handle our entire lives as an artistic discipline. Ikebana allows us to develop discipline and it shows how much general appreciation we have developed, and how much we have learned a sense of being in the world harmoniously. Ikebana discipline is not just arranging pretty flowers, making them into a beautiful arrange-ment. More fundamentally, it is a reflection of oneself."[9]

In everyday life one could also realize every action as complete, like placing a flower or a single breath. When you say hello, shake hands, lift a glass of wine, start your car, put the kettle on, each action has its beginning, middle, and end, each is complete. When you paint, there is first the empty canvas, then the first brush stroke, then the completion of the painting. When you write a letter, there is an empty piece of paper. Then you write, "Dear Friend," then you write your letter, and finally, "Yours sincerely," and it is finished. Such appreciation of the present moment is known as "nowness"—the realization that, at each moment, this very moment is the only occasion of your life, uncorrupted by past or future.

According to the teaching of Shambhala, this is the basis of how to live life in accord with unconditioned nature. Every thought, every word, every gesture could be complete, and take its natural place, appropriate, harmonious, and elegant. This possibility is based on the principles of *shamatha* and *vipashyana*. The heart of the practice is synchronizing or joining mind and body. Or, to say the same thing, it is joining awareness and perception.

The process of joining mind and body is a process of continually letting go. At each step of the journey, each stage of mindfulness-awareness practice, we discover the central conception of "I," the fear and aggression which come from holding on to it, and all the other conceptions, personal views, and opinions which surround it. Thus at each stage we let go of this conjecture and its surrounding conceptual structures in order to see clearly beyond it. Out of this a natural ethic begins to develop, of recognizing the unconditioned goodness of others as well as oneself and of letting go of one's personal discovery of goodness in order to promote the fundamental welfare of others as well as oneself.

Synchronizing mind and body in nowness is seeing the world directly, beyond language. As Trungpa says: "Sometimes when we perceive the world, we perceive without language. We perceive spontaneously with a pre-language system. But sometimes when we perceive the world we think a word first, and then we perceive. In other words, the first instance is directly feeling or perceiving the universe, the second instance is talking ourselves into seeing our universe. So either you see the world through the filter of your thoughts, by talking to yourself, or you look and see beyond language—as first perception . . . . When you feel that you can afford to relax and perceive the world directly, then your vision can expand. You can see on the spot with wakefulness. Your eyes begin to open, wider and wider, and you see that the world is colorful and fresh and so precise; every sharp angle is fantastic."[10]

17 Society

I began this book by pointing to the situation throughout the world today and to various proposals that have been made to speak to this "turning point." I suggested that so long as such proposals, no matter how well intentioned, are based on unexamined conditioning and beliefs, including especially beliefs in "self," then they can only breed further struggle and conflict. I introduced the idea of unconditioned nature and the possibility of training perceptions to realize their unconditioned basis, and this has been the theme of the book. Our analysis of the profound involvement of the idea of "self" at every level of perception pointed to the fact that such training is in no way self-centered. It is, rather, the only way fundamentally to overcome the distortion and aggression of self-centeredness and actually to build a good society. As I have said, the training of perception in meditation practice is simply practice: transforming perception leads to transforming action. This action can then be fundamentally based on gentleness, nonaggression, and overcoming fear through the courage of being genuine. This can form the basis for enlightened society.

Role of Beliefs and Fear in Society

Psychiatrist Roger Walsh, in a penetrating analysis, has concluded that the problems that the world faces today, problems of hunger, poverty, overcrowding, and the threat of nuclear catastrophe, are all of psychological rather than physical origin.[1] There are enough physical resources and technical knowledge to overcome each of these problems at this time. That the world seems to be frozen into dangerous inaction stems from psychological problems which a psychiatrist meets every day in his office and which are reflected on a social and international level. Of the problems that Walsh focuses on, the most basic are cognitive factors of beliefs and dissonance, defense mechanisms, and the central role of fear. Each of these problems has appeared in our discussion in previous chapters as ways of trying to avoid or cover over our unconditioned nature.

Of beliefs and presuppositions, Walsh says, "Beliefs tend to modify what we look for, what we recognize, how we interpret and how we respond to those interpretations. Yet what is absolutely crucial about these processes is that they tend to be self-fulfilling and self-prophetic yet largely unconscious." He points out the extraordinary difficulty of becoming aware of our own presuppositions, then continues: "But become aware of our presuppositions we must, if we are to recognize the unskillful beliefs which are directing us both individually and culturally towards our contemporary crises." He points to beliefs such as "My beliefs are the truth and the only truth," "It's their fault not ours," "There's nothing that can be done," as examples which occur on the individual level and also have dangerous and powerful consequences on the social level.

Of defense mechanisms, Walsh comments, "From the psychodynamic perspective, defense mechanisms constitute the heart of individual psychopathology. As at the individual level, so also at the social; defense mechanisms offer rich insights into many facets of our contemporary dilemmas. The reasons for this become apparent if we remember that many of our current difficulties stem from our lack of awareness of their true nature, and that defense mechanisms operate by reducing and/or distorting awareness. One of the persistent sources of despair to those working in these areas is the recognition of just how hard it is to reach or sustain awareness of the true state of the world." He suggests that "we wish to deny not only the state of the world but also our role in producing it." Thus we project the concept of "enemy" onto others, and likewise others project the same concept onto us.

The central role of fear is put thus by Walsh: "When these defenses, distortions, addictions and aversions are examined, it can be seen that they represent unskillful attempts to deal with fear. Indeed from this perspective the current international and nuclear threats can be seen as expressions of fear: fear of attack, fear for our survival, fear of losing our comforts, lifestyles, ideologies and economic supplies."

Individual Training is the Basis of Enlightened Society

We have seen all these factors—beliefs, defense mechanisms, and fear—operating at a deep level of perception in the individual, and the force of Walsh's analysis is to show that these same factors extend beyond the individual to the entire culture. Nevertheless, because culture itself is rooted in individuals, the problems arise at the individual level and radiate out to society, to influence other individuals. Likewise to resolve these conflicts and change cultural attitudes, we must work at the individual level. It may be helpful as a beginning to try to change cultural attitudes, to replace commonly held "negative" beliefs by "positive" ones. Indeed, it is important to do so. However just as it is at the individual level, so it is at the social level, that a shift from one set of beliefs to another is still only a shift of beliefs and does not bring about the fundamental change of freedom from attachment to belief as such. A fundamental change in society can happen only if that society begins to be based on total freedom from partiality and exclusivity, that freedom which comes from recognizing the unconditioned goodness of human nature. Although such impartiality, such freedom from the need to believe, may not be fully attainable by all individuals in one lifetime, the *aspiration* toward such freedom already makes a profound difference. The acknowledgment of the unconditioned aspect of human nature, which translates experientially as the freedom from the conjecture of a "self," the discovery of this freedom, at first in occasional glimpses, and the aspiration to fully realize it, are the foundation of genuine change in society.

As we have seen throughout the book, recognition of unconditioned goodness is a thoroughly practical proposal in that the skillful means are available to accomplish it and have been taught and experienced for many generations. I refer of course to the practice of sitting meditation and other related contemplative practices such as ikebana, t'ai chi chuan, and so on. However, in order to be effective in transforming society, these techniques must be embedded within a context, a tradition, or a teaching which complements the experiential discoveries of the practice with intellectual understanding. In this way, individual discovery becomes available across society and from generation to generation. In this way, also, the realization of unconditioned nature can enliven and transform the structures, activities, and organization of the larger society.

Two such traditions which have provided a foundation for spiritual training and social organization for more than two thousand years are the Buddhist and neo-Confucian traditions. Ironically, both these traditions have been thoroughly misrepresented in Western literature, both scholarly and popular, until the past half century or so. Both represent a complete philosophical and practical way of living ordinary life as sacred in the sense of appreciating the fundamental purity and goodness of the world. Yet each has been misrepresented in almost contrary ways. Buddhism has been classified as one of the "world's great religions," while it is by no means a "religion" in the conventional sense of worship of a deity, striving to escape from this world, and so on. On the other hand, neo-Confucianism has been represented as a secular, humanistic doctrine of social responsibility and government, while it is by no means merely "secular" in the conventional sense of denying the unlimited possibility for men and women to transcend their self-preoccupation through a life of training.

Two Traditional Teachings of Unconditioned Goodness

The Buddhist tradition began in the sixth century B.C., when the Prince Gautama at age twenty-nine left the luxury and security of his father's palace to seek an understanding of the nature of human life. After many years of study, and suffering through many ascetic and yogic practices, the prince sat down one night beneath a tree (later named the Bodhi tree) and resolved not to arise until he had understood and realized his essential nature. As dawn broke, Prince Gautama fully realized the unconditioned basis of all existence. This is referred to as his moment of awakening, and from then on he was called the Buddha, meaning "he who is awake" to his true nature. The Buddha at first remained silent for seven weeks, unsure how to communicate to others his ineffable realization. Nevertheless, he began to teach and instruct others in how to attain such realization and continued to teach for the remaining forty-five years of his life. The instructions of the Buddha, now known as Buddhism, spread from India to China, Japan, Tibet, Korea, Sri Lanka, and many other countries. As it spread, it took on the cultural and philosophical forms of each country. Thus, Buddhism in Japan appears to be very different from the Buddhism of Sri Lanka or of Tibet. Nevertheless, the basic message remains the same: first, recognizing the suffering caused by ego-centered clinging; second, how to let go of clinging and uncover our unconditioned nature of goodness through the practice of sitting meditation; third, how to manifest the unconditioned goodness of all in compassionate action to alleviate the suffering of others.

The teachings of Shambhala began in the kingdom of Shambhala, which may have existed in central Asia between Mongolia and the Sinkiang province of China. Legends tell us that the first king of Shambhala, Dawa Sangpo, invited the Buddha to Shambhala to teach. King Dawa Sangpo told the Buddha that he wanted his subjects to receive not a religious or monastic training, but instruction on how to live wakefully and completely in their everyday lives. Accordingly, the Buddha taught the Kalachakra Tantra of Vajrayana Buddhism, which is now regarded as one of the most profound teachings of Buddhism.

The subjects of the kingdom of Shambhala began to practice meditation and form their society according to these instructions. Thus the Shambhala kingdom became a model for an enlightened society, that is, a society based on the recognition of the basic nature of unconditioned goodness and gentleness of human beings and the aspiration of all members of the society to train in order to manifest such gentleness fearlessly in the world. We have met with other key doctrines of Buddhism as they were relevant at various points in the book, so we do not need to go further into the Buddhist tradition here.[2]

Confucius recognized the extent to which the human heart is guided and formed by the social relationships of which we are inevitably a part. In fact, for Confucius the individual is *defined* by his relationships and by the proper fulfilling of his responsibilities in those relationships—there simply is no individual "self" whose desires have to be constantly satisfied in isolation from the society in which it has its being. For Confucius, then, the realization of sagehood—the manifestation of unconditioned goodness, humaneness—consisted in the genuine, wholehearted fulfillment of the responsibilities and rituals of the society into which one is born. This may at first seem paradoxical: how can we realize our unconditioned nature by carrying out prescribed forms? But it is paradoxical only for the modern mind for which freedom consists, par excellence, in not having to do what "I" don't want to do, or being unrestricted in what "I" do want to do. For Confucius, this ordinary world is itself sacred and, through genuine sincerity and respect for the sacredness of this world, the individual can learn its functioning and the sacredness of his own place in it. Thus gradually he uncovers his unconditioned mind or "heaven nature," which is not separate from the sacredness of the ordinary world. Confucius recognized that almost every human act is a social act, and therefore a form of ritual, *li*. From saying hello and shaking hands, to serving and eating food, all the way up to the conduct of government, all is ritual, handed down from generation to generation. As Fingarette says, "Confucius may be taken to imply that the human being has ultimate dignity, sacred dignity by virtue of his role in rite, in ceremony, in *li*. We must recall that Confucius expanded the sense of the word *li* to envision society itself."[3]

The life and writings of Confucius himself are rather obscure, and it was a student of his teaching who lived one hundred years later who was to systematize Confucian teachings and start them on the way to becoming the underpinnings of Chinese culture. This student, Mencius (371–289 B.C.), was the first to explicitly teach the fundamental goodness of human nature: "If you let people follow their feelings

[original nature] they will be able to do good. This is what is meant by
saying the human nature is good. If man does evil it is not the fault of
his natural endowment. The feeling of commiseration is found in all
men . . . [and] is what we call humanity. Humanity [is] not drilled
into us from outside. We originally have [it] with us. Only we do not
think [to find it]. Therefore it is said, 'Seek and you will find it, neglect
and you will lose it.'"[4]

Almost contemporaneous with Mencius was another Confucian
commentator, Hsun Tzu (298–238 B.C.), who proclaimed the view that
human nature is evil and has to be trained to overcome this evil,
although he recognized an almost unlimited capacity of human nature
to be so trained.[5] (For Mencius this capacity is itself an aspect of the
inherent goodness.) This debate between followers of Mencius and
followers of Hsun Tzu continued over the centuries, and I mention it
here since a similar debate has also, quite obviously, raged over the
centuries in the Western tradition as well. Later writers realized that
there is a difference of perspective but not a fundamental difference
between these viewpoints, that while Mencius was focusing on the
fundamental, unconditioned nature, Hsun Tzu was focusing on the
defilements or obscurations of that unconditioned goodness, namely,
our conditioning and beliefs. This is brought out by a thirteenth-cen-
tury neo-Confucian, Tai Chih: "People talk about human nature—
some say it is good, others that it is bad. Generally they prefer
Mencius' view and reject Hsun Tzu's. After studying both books I
realized that Mencius is talking about the heaven [unconditioned]
-nature and what he calls goodness refers to its uprightness and
greatness. He wished to encourage it . . . . But Hsun Tzu is talking
about matter [conditioned] -nature, and what he calls the badness of
human nature refers to its wrongness and roughness. He wished to
repair and control it . . . . Thus Mencius's teaching is to strengthen
what is already pure, so that defilement tends to disappear of itself.
While Hsun Tzu's teaching is to remove the defilement activity."[6] It is
Mencius's point of view that is more fundamental, since Hsun Tzu is
able to maintain his only by ignoring the unconditioned nature or
trying to aggrandize it under man's control which is not possible.

Confucianism weakened when Buddhism entered China in the
first century of this millennium and vied with Taoism as the ruling
philosophy for the next eight centuries. In the ninth century,
Confucianists responded by combining the ethical-social principles of
early Confucianism with the cosmology and natural science of Taoism
and the studies of mind and metaphysics of Buddhism to produce a
new synthesis known as neo-Confucianism. In this new synthesis,

Mencius's view of human nature was dominant, and the training process of sitting meditation played a major role.[7]

Neo-Confucianism became the foundation of Chinese government and culture from that time until the nineteenth century. The Confucian form of government was a hierarchy stretching down from the emperor. A member of government was known as a *shih* or warrior-gentleman, and to be a *shih* was regarded as the highest attainment for a young man. Selection was by examination; therefore the hierarchy was based on merit rather than on birth. The *shih* was examined in every aspect of life: his understanding of Confucian classics; his proficiency in the martial arts, horsemanship, archery, swordsmanship; his accomplishments in calligraphy, painting, and poetry; and finally his level of understanding of sagehood. Thus the hierarchy was spiritual as well as secular, and the emperor was regarded as an accomplished sage. Historically, of course, each era only more or less came up to this ideal, depending on the forthrightness and genuineness of the *shih* and on the genuine spiritual attainment of the emperor and his actions in benefiting the people. There were glorious times in Chinese imperial history as well as times of decay.[8]

In later centuries, Buddhism and Confucianism were inextricably interwoven in the government and culture of China, Korea, and Japan. This cooperative blend forms the basis of the martial, or contemplative, arts of Japan: kendo (swordsmanship), kyudo (archery), aikido (combat), chanoyu (tea ceremony), ikebana (flower arranging). As we have seen, each of these disciplines provides a form through which one's state of mind-heart (*kokoro*) is revealed through one's action. Conversely, each provides a path of spiritual training— one's mind becomes clarified and awakened as one refines one's action, whether it be handling a sword, serving tea, or placing a flower.

We should not leave the impression that the Confucian search for sagehood was merely an attempt to unreflectingly conform to the norms of a society no matter how corrupt or decayed. To understand this, we need to understand the notion of *tien* or heaven.[9] Heaven in Confucianism is the unconditioned nature of all things, the essence and fundamental norm of the way things are.

Heaven, Earth, and Man

Heaven is the clear blue sky above us, and the open space surrounding us. Heaven also refers to a sense of vastness, primordialness, before thought, before concept, before form, before time and space altogether. Heaven is vast, unbounded, empty in the sense of empty of conceptualization, and full in the sense of full of potentiality. It is the unconditioned aspect of the totality of being and nonbeing, which I presented in Chapter 15.

Heaven is the aspect of primordial mind, primordial intelligence, beyond thought and concept. Therefore heaven is beyond hope and fear. It is a sense of purity, clarity, and brilliance that accommodates everything. It is beyond hope of continuing and fear of not continuing. It is absence of struggle, absence of burden or problem. Therefore, there is a quality of relief, openness, letting go without dwelling and holding on. So in the space of heaven which accommodates all possibilities, there is tremendous freedom without plan or preconception. Because there is such freedom, there is in the ordinary world a sense of sacredness. This sacredness does not come from something else; it is not conditioned by someone else's promise, or by any promise at all. Simply, everything in its own nature is sacred, because of its fundamental unconditioned nature, its heaven aspect. Heaven brings the profundity and vastness of perception, which we discussed in Chapter 15.

Earth is the real, solid earth beneath our feet. Earth is also the reflection and manifestation of the potentialities of heaven, the actual bringing into fact of the many possibilities envisioned in heaven. Heaven gives the possibility of choice, of norms, of inherent natural order. Earth is the actualization of those norms.

The man principle is that which joins heaven and earth. While the essential nature of man is unconditioned goodness, the constitution of man is conditioned, and his function is to be a clear, unobstructed channel bringing together the unconditioned and the conditioned. When confusion is clarified and obstacles are removed, heaven and earth are naturally joined. Man joins heaven and earth, therefore, by simply being genuinely who he is. "The greatest virtue of Heaven and Earth is to live."

The search for sagehood is the search to uncover and follow the way things are, the way of heaven. The unique role of man, according to Confucianism, is to join heaven and earth, that is, to bring the unconditioned nature of all things into the practicality of day-to-day affairs and social relationships: this is the way of the sage. According to the tradition, the gloriousness or decay of the kingdom depended on the extent to which the emperor—and, following his example, all the subjects—adhered to the path of sagehood, the way of heaven. When the emperor followed the way of heaven, the kingdom flourished; when he did not, it fell into disorder.

The founders of the Confucian tradition differed on many points, but nevertheless, they "had a common spirit," according to Charles Hucker.[10] This common spirit "can be expressed without too much distortion in seven propositions:

1. The universe and mankind are governed by an impersonal but willful Heaven (*t'ien*).
2. Heaven wills that men be happy and orderly in accord with cosmic harmony (*tao*).
3. An ethical and virtuous life is the appropriate human contribution to the cosmic harmony.
4. Virtue (*te*) is developed and manifested in proper conduct (*li*). It should be specially noted in this regard that virtue is not something nurtured and secluded in one's secret self; man is only what he does. As Mencius said, 'If you wear the clothes of [the sage-ruler] Yao, speak the words of Yao, and carry out the actions of Yao, you will then be a Yao. But if you wear the clothes of [the Hsia dynasty tyrant] Chieh, speak the words of Chieh, and carry out the actions of Chieh, you will then be a Chieh.'
5. In important particular crises it is not easy to know Heaven's will or to choose between conflicting values: this is especially the case for a ruler.
6. Proper conduct, especially in important matters, often requires wisdom or sageliness; and this can be acquired only through sober study of the precepts of prior sages and the lessons of history, and through earnest effort to cultivate oneself accordingly.
7. Individual men and human society are perfectible; that is, every man potentially can be a sage, and society potentially can be made harmonious and fulfilling for all."

Thus we can see the complementarity of the Buddhist and neo-Confucian traditions. The Buddhist tradition proceeded by training the individual practitioner to see through his or her self-obsession and to work for the benefit of others. Such a practitioner was then able to adopt the forms of relationship of established society and use them to propagate wakefulness and to help others. Neo-Confucianism proceeded by organizing society on spiritual principles that already fundamentally avoided the error of promoting individual, isolated "selfhood." Therefore, by following the norms of such a society and finding his place in it, the warrior-gentleman could himself see through this error and attain sagehood. We are not suggesting, of course, that in either case the life journey of a person was automatically guaranteed to produce realization without effort, or that society was guaranteed to be free from corruption. All this depends on individual effort.

I have described Buddhism and Confucianism as two historical teachings of spiritual training which can form the basis of enlightened society. The reader may wonder why I have not also mentioned Christianity in this context. Certainly Christianity has been a major civilizing influence in Western history along with science. However, I have been particularly emphasizing the importance of recognizing the unconditioned aspect of the world, as well as the importance of the actual practice of contemplative disciplines. It would take us too far afield here to discuss the role of the unconditioned in contemplative Christianity and how this has, in turn, influenced Christian society, but such discussion would be well worthwhile. Dialogues between Christianity and Buddhism over the past few decades have disclosed that the contemplative trends of Christianity are very much alive and reappearing after perhaps many centuries of being unacknowledged.

18   The Power of the Ordinary World

In the previous chapter we saw that Buddhism and Confucianism are two great traditions that have been available for two thousand years as a foundation for building an enlightened society based on the recognition of unconditioned goodness, and on the individual spiritual journey of awakening. We may now ask, what is available in the modern world for someone wishing to take such a journey?

Evaluating Spiritual Teachings

To try to answer these questions, one has to be quite careful. There is a very large range of disciplines, paths, and so-called "new religions" being offered currently in the West: from those that are obviously destructive of human dignity and intelligence with no redeeming features, such as Jonestown and Synanon, to those that just as clearly have basically positive characteristics, such as the traditional Buddhist or Vedantic Hindu teachings. Along with these are various popular movements which may be highly questionable but not obviously destructive, whose ambiguity makes them in some sense even more potentially harmful. Scientology and Moon's Unification Church may be examples of the latter type.

Western sociologists, especially those who regard personal experience of a discipline as somehow a liability to their "objectivity" and therefore have at best secondhand experience of what they are talking about, have tended to thoroughly confuse the issue. Not recognizing the actuality of spiritual practice and transformation, that is, of nondualistic understanding which is neither "objective" nor "subjective," they have simply lumped the entire range into one meaningless category. On the other hand, some sociologists who *have* recognized the actuality of human spirituality have, in an effort to be ecumenical, taken the approach that all disciplines are saying the same thing—the "perennial philosophy" approach—which leads to another relatively worthless lump. Morris Berman, for example, generally acknowledges the possibility of spiritual transformation and argues cogently for it throughout his book, *The Reenchantment of the World*. Yet, in the final chapter he speaks of the dangers of "guruism," by which he appears to mean devotion to the personality of a teacher rather than to his basic sanity and to the teachings he embodies. However, he makes no distinction between these, and therefore seems not to appreciate the importance of the lineage of guides and teachers in any spiritual path of training. At the same time, warnings such as Berman's should not be overlooked, for a great deal of distortion and misrepresentation of authentic teachings has undoubtedly happened. There is always a possibility of using spiritual practice as a means to greater egoism, rather than to greater opening and kindness to others.

How, then, are we to assess the situation? Are there criteria which may be generally applied to make some useful distinctions between the various movements? A step in this direction has been taken by sociologists Dick Anthony and Tom Robbins in their work, which is the result of an extensive and in-depth survey including lengthy taped interviews with participating members of certain groups. Anthony and Robbins propose a threefold classification: monist/dualist; technical/charismatic and one-level/two-level.

Monist/dualist refers to the question of whether all people or only a particular group, the "select," are said to be able to attain the fruition of the teachings. Monist types of teaching are accessible to all, while dualist teachings, in various ways, exclude most participants from the highest possibilities. Charismatic movements focus attention on the personal power of the group leader, while technical movements emphasize the techniques and practices that will aid the student. One-level teachings focus on a purely secular and often materialistic attainment, while two-level teachings acknowledge in some manner transcendent or sacred possibilities. On the whole, Robbins and Anthony found that dualistic, charismatic, one-level groups tend to be the most negative or problematic. The most positive and nonproblematic groups were the monistic, technical and two-level groups. Examples of the first extreme are Jonestown, Synanon and followers of Reverend Moon, while examples of the latter groups are Integral Yoga, Hindu Vedanta and Buddhist traditions.

In *Eye To Eye*, Ken Wilber takes a step further this attempt to understand on a rational level the ground of healthiness in a spiritual or religious teaching. I will very briefly summarize Wilber's arguments. Wilber offers three categories within each of which various teachings may be distinguished as being problematic or nonproblematic. These categories are: (1) authenticity, (2) legitimacy, and (3) authority. By authenticity, Wilber refers to the degree to which a teaching recognizes a larger scope to human existence beyond egoism or personal power. An authentic teaching is one which is able to uplift people beyond the ego-centered, merely rational phase of existence to a phase which is transrational and genuinely intuitive. Wilber very helpfully distinguishes between teachings that are *trans*rational and those which are *pre*rational. The confusion between these he calls the pre/trans fallacy. A transrational teaching goes beyond commonsense ego-centered rationality, without negating rationality as such, while a teaching which is prerational denies rationality and tries to return the student to the prerational phase of a child or of an earlier mythic phase of human development. Confusion between these two leads to the typical error made by Suzanne Langer, for example, and reiterated by Berman, of considering that all meditative techniques have the problem of denying thought. On the other side, such confusion led to the distortion of Zen in the sixties, in which all thought was regarded as "fucked up" and there was a call to return to impulse. However, when pre- and transrational are distinguished, it becomes fairly easy to see which teachings are authentic in Wilber's sense.[1]

By legitimacy, Wilber refers to the way in which a teaching validates itself. That is, says Wilber, "we are looking for the degree of integration, meaningful engagement, purpose and stability . . . . Is legitimacy conferred by a whole society or by a tradition or by a single person?" A problematic teaching is one which derives its legitimacy from the activity of a single person, rather than from the larger community or a lineage or tradition.

Finally, in discussing authority, Wilber distinguishes two aspects of benign and healthy kinds of authority. First, functional authority: authority which is conferred on an individual by the function he or she performs. For example, a school teacher expects a child to do arithmetic the way the teacher tells him. This is simply the only way the child will learn arithmetic from that particular teacher. When the child has mastered arithmetic the way the teacher taught him, he may then be able to find better ways of doing it; a genuine teacher will acknowledge this. The second aspect of healthy authority is that it should be "phase-specific": that is, the teacher must, temporarily, be authority for the student as long as the student has not fully accomplished what has been taught. However, when the student *has* accomplished that phase of the teaching, then the teacher ceases to be an authority. Furthermore, even here the notion of authority is as a guide who, having already traversed a part of the journey, can show the student how to discover it for himself or herself. The teacher always regards himself as a student; there is always further learning possible. Therefore, a teacher can never become an absolute or permanent authority. These, then, are Wilber's three criteria for assessing a path of training.

I would like, in concluding this book, to describe briefly two teachings which are available now and which provide an up-to-date nonreligious context for a path of meditative training. I refer to the Fourth Way and the Shambhala teachings. There are, of course, other such teachings available, and I present these two only because they are the two with which I am personally familiar and know to be genuine and non-problematic, in the sense of Wilber's three criteria. In addition to these two nonreligious teachings, there are the teachings of the great tradition of Buddhism itself, which are now quite widely available in the West. Some may be deterred from investigating Buddhism because of its religious connotations. But, as I have indicated, Buddhism may be taken not as a belief-system but as a training in how to live one's life fully, without reliance on narrow beliefs.

Both the Fourth Way and the Shambhala teachings are modern presentations, for a worldwide community, of teachings which had their origin many hundreds of generations ago in central Asia. In the early decades of this century, there were various attempts to adapt the doctrines of the "mystical East" to Western conditions. In this case, the term "mystical" seems to represent the very partial and sometimes absurd understanding of Western explorers and storytellers more than anything inherent in the doctrines themselves. The "secret" and "occult" societies which formed in order to propagate these adapted doctrines were largely based on fantasy and wish fulfillment: from the romance of Shangri-la, to the multiple planes of existence of the occultists, to the "magic" of Alistair Crowley. Even apparently serious writers such as Madame Alexandra David-Neel, author of *Initiations and Initiates in Tibet*,[2] have been drawn into this dreamworld to some extent. Unfortunately, this misrepresentation is perpetuated by the popular science press, in which statements of the doctrines are taken out of their proper context: the context of a lifetime's training in meditation. Omitted is the fact that for the teachers of these doctrines, the training itself is the important point. They warn novices *against* focusing on such goal-oriented statements, which are intended only as metaphors for direct perception and as lures to encourage the students on a difficult path of individual spiritual growth. It is the training itself, not the goal, that is regarded as important in these traditions.

The Fourth Way

A powerful and genuine effort to make methods of spiritual training available for modern times was that of G. I. Gurdjieff in the first half of the century. Gurdjieff was born in Russia around 1877.[3] As a young man, he traveled in India, Tibet, and the Middle East, "in search of the miraculous," that is, in search of that which brings out all the possibilities for human life. He appeared again in Moscow in 1912, and for the next thirty-six years he taught, first in Russia, then in France and America. He called his teaching the Fourth Way to distinguish it from three traditional approaches to spirituality: the "way of the yogi," involving extreme development of intellect; the "way of the monk," involving extreme development of devotion of the heart; and the "way of the ascetic," involving extreme development of physical dexterity. The Fourth Way, the way of the householder, is a method of training using the experiences of everyday life to observe and experiment with the energies of the human body, mind, and emotion. The aim is the harmonious integration of these three. The Fourth Way appears to be a blend of Sufism, Buddhism, and Christianity, which Gurdjieff sometimes referred to as "esoteric Christianity."[4]

The Fourth Way points out that while everyone thinks of himself or herself as a single unit, a self, an "I," we usually find our identity, our "I," in every fleeting thought, feeling, or action that comes along. Our identity changes from moment to moment, and often one thought or feeling is altogether in conflict with the immediately preceding one. We flare up with anger or passion without knowing how or why, and say "I am angry" or "I am in love." We feel depressed or elated in response to changing circumstances and identify our "I" with these moods. In particular, there are three major functional centers in the human being. These are the thinking center, the emotional center, and the moving center. These three each have a separate "I" and are very often in conflict with one another: we think something that we do not feel; we feel desire for something that our body rejects; our body acts against what we think we should do, and so on. Altogether, according to Gurdjieff, our life is generally in rather a mess.

In pointing out this mess, Gurdjieff is asking men and women to cease taking their lives for granted and to begin to question profoundly the whole basis of their existence. Men and women are asleep, hypnotized by their own thoughts, spinning helplessly from moment to moment. Only when they begin to see this and question it can they even realize that they are asleep. And realizing that one is asleep is the only way to begin to wake up.

The "effort" of the Fourth Way consists of "self-observation," that is, of simply, nonjudgmentally noting at various moments throughout the day one's states of thought, feeling, and body. Self-observation is said to be like taking random snapshots of oneself and gradually thereby beginning to realize one's fragmentation and disorder. Various still and moving exercises are provided in order to facilitate such self-observation. In particular, the day begins with a short sitting exercise known as "collection," which is very similar to mindfulness practice—simply noting feeling, thought, and especially physical sensations. In addition to providing training in nonjudgmental noting, or self-observation, these exercises also provide the possibility for a moment in which body, emotion, and thoughts are harmonized. At such a moment, an entirely new perception or quality of attention is said to arise, and with it an entirely different sense of who one is. It is clear that self-observation is essentially the practice of mindfulness leading, at first only occasionally, to the harmonizing of thought, emotion, and body. I have described this earlier as joining mind (in the larger sense) and body.

Gurdjieff distinguishes between "essence" and "personality." Personality refers to what we think we are. It is the sense of a single "self" that has been built up in us by the various forces of external circumstances—the expectation of our parents, our teachers, our friends—together with our internal response of fear and imagination, which we call "consciousness." All this is glued together, by our refusal to look at who we are, into a solid fixed thing, our "I," our personality which we spend our lives trying to sustain. At the same time, personality is also the best of what our upbringing has given us, of what we have managed to become through culture and education.

Essence is our inherent nature, the energies and particular qualities of our moving, emotional, and thinking centers. It is the possibilities for both positive and negative manifestation that we are born with. Another way of speaking about the need for bringing harmony to our being is therefore in terms of the harmonizing of essence and personality. Very often our upbringing, rather than nourish the genuine nature of our essence, simply covers it over with a veneer. The essence remains a baby. Self-observation then begins to soften the hard shell of personality so that what is dead may fall away and what is nourishing may become food for essence.

The heart of our essence, and therefore the basis of our whole being, is what Gurdjieff calls our "nothingness." Gurdjieff is said to have remarked that "only when a man begins to know his own nothingness can he really be helped." That is to say, only when a man or woman has begun to recognize his or her unconditioned nature can there be a possibility of being fully human.

The spiritual path is often taken to be solemn and serious, yet cheerfulness and good humor seem to be the heart of it. Gurdjieff himself is said to have had a remarkable sense of humor, taking delight in joking with his students and poking fun at their self-solemnity. During the last years of his life he used to gather his students in his small apartment in Paris for magnificent meals of spicy food and Armagnac brandy. These meals always concluded with a round of "toasting the idiots," a recognition perhaps of both the insanity of human dreams and the wisdom and humor that can begin to awaken us when we see our "idiocy." Gurdjieff is reported to have remarked to his wife about his students, "All I want for them is that they be happy." By "happy" he meant not the superficial happiness which is contrasted with depression, but the fundamental sense of well-being that comes from being fully human.

Working together in groups, forming communities based on these principles, and extending out to being active in society are all important aspects of the Fourth Way. There are large numbers of such groups throughout the world today, working to bring about a society based on genuinely knowing who we are and on the possibilities for awakening.

Unfortunately, when Gurdjieff died he did not leave behind a clear line of succession. Many schools arose in his name, some genuine, some imitations, and some slipping back into occultism. Nevertheless, Gurdjieff's Fourth Way continues to be one of the oldest and most vital attempts to join together various Eastern and Western traditions of spiritual training and to adapt their form for modern understanding.

Shambhala: The Sacred Path of the Warrior

An up-to-date, nonreligious context for the practice of meditation has begun to develop in the past decade under the guidance of Chögyam Trungpa XI. This path of training is known as Shambhala Training. Chögyam Trungpa XI was born in Tibet and recognized at an early age as the reincarnation of the Tenth Trungpa. At the age of five, his formal training began in the scholarly and meditative disciplines of the Kagyü and Nyingma Buddhist schools—the two major contemplative schools of the Tibetan Vajrayana Buddhist tradition. He was empowered as holder of these lineages and invested as supreme abbot of the Surmang Monasteries and civil governor of the Surmang region of eastern Tibet. He escaped from Tibet in 1959 after the Chinese invasion. After residing in India for three years, he attended Oxford University as a Spaulding scholar, studying the Western traditions of psychology, comparative religions, and the fine arts. It was here also that his lasting appreciation of the culture of Japan began and he studied ikebana with the Sogetsu school.

Chögyam Trungpa began to teach Buddhism to Western students while in Great Britain and founded a meditation center, Samye Ling, in Scotland. A turning point in his relationship to Western society came after a serious car accident in 1969. As he says in his autobiography, *Born in Tibet*, "This [car crash] led to me taking off my robes. The purpose was for me personally to find the strength to continue teaching by unmasking, and also to do away with the 'exotic' externals which were too fascinating to students in the West."[5] He married an Englishwoman later in 1969 and moved to the United States, where he began once more to teach Buddhadharma. Recognizing the need for a nonreligious presentation and for a cultural container for the Buddhist teachings, and also recognizing the close similarity between neo-Confucianism and the cultural and governmental training he had received in Tibet based on the tradition of Shambhala, Chögyam Trungpa began to present the teachings of Shambhala in 1976. He is also the founder and president of Naropa Institute, a contemplative liberal arts college whose goal is to join intellect and intuition in a traditional university setting.

Chögyam Trungpa is a man of very great compassion, with the extraordinary combination of having vast and profound mind yet being thoroughly down to earth. He loves to share jokes with his students and frequently reminds them that a natural sense of humor is the best antidote to ego's tightness and fear. His senior Western student, and chosen successor in the Vajrayana Buddhist lineage, is an American, Thomas F. Rich, now known as Ösel Tendzin, who is co-founder, with Trungpa Rinpoche, of the Shambhala Training Program, through which the Shambhala teachings are introduced to the public. The way of Shambhala offers a genuine and practical path through which unconditioned and conditioned natures, heaven and earth, may be joined in individual lives and in society.[6]

The Shambhala teachings draw to some extent on the teachings of both the Buddhist and the neo-Confucian traditions. We saw that Confucianism and Buddhism are two great contemplative traditions that have always acknowledged the unconditioned basis of man's nature. For each of them, the journey of a human being in everyday life is to bring the unconditioned and the conditioned together, to join heaven and earth. This journey is being presented for modern times as the journey of warriorship in the teachings of Shambhala. The teachings of Shambhala have also been called the Path of the Warrior. "Warrior" here does not mean one who makes war, but rather one who overcomes war and personal aggression, one who is gentle and fearless.

There are groups of practitioners of the Shambhala teachings throughout North America and in Europe. Such practitioners train together in intensive training programs studying the principles of Shambhala and mindfulness-awareness practice, (as we described in Chapter 16), as well as other contemplative practices. Their professions range from shopkeepers and carpenters to housewives, lawyers, scientists, and doctors. As with the Fourth Way, so with the Shambhala training programs; the aim is a very practical one: to try to bring the principles and practices of the training into every aspect of everyday life, and therefore gradually to begin to form a society based on these principles.

The teaching of Shambhala is a statement of secular enlightenment, that we can bring about fundamental change in our lives without dependence on religious doctrines. However, it is not merely another program for "self-improvement." Shambhala is very much concerned about our responsibility to the rest of the world. We could say that the primary purpose of the vision of Shambhala is to "encourage a wholesome existence for oneself and others." It is a statement that there *is* basic human goodness and wisdom which is capable of uplifting this world. This goodness and this wisdom begins with recognition of who we are, and with not being afraid to be who we are. Such fearlessness comes not from making a big deal out of one's life or pretending to be what one thinks one should be, but from being simple and genuine.

In discussing fearlessness, Chögyam Trungpa speaks of the "genuine heart of sadness." Sadness here does not mean depression, self-pity, or sentimentality. It refers more to a feeling of youthful softness and tenderness. There is also a touch of longing to communicate which comes from being so full. The word *sad* has the same root as the word *satisfy* and contains in it a sense of aloneness and of the fullness of maturity. Watching a flock of birds flying across the autumn sky; leaving home as an adult, maybe never to return; finishing a meal with old, good friends; hearing a familiar song sung beautifully in the distance when you are far from home: in all these, joy and sorrow are combined, and cannot be fully communicated. Yet we may wish so much to share our joy and sorrow with others.

"Sadness hits you in your heart, and your body produces a tear. Before you cry, there is a feeling in your chest and then, after that, you produce tears in your eyes. You are about to produce rain or a waterfall in your eyes and you feel sad and lonely, and perhaps romantic at the same time. That is the first tip of fearlessness, and the first sign of real warriorship. You might think that, when you experience fearlessness, you will hear the opening to Beethoven's Fifth Symphony or see a great explosion in the sky, but it doesn't happen that way. In the Shambhala tradition, discovering fearlessness comes from working with the softness of the human heart."

The Great Eastern Sun

In the Shambhala teaching, the image of the Great Eastern Sun is used as a symbol of the vision of celebrating life and the sacredness of the world, and this is contrasted with the image of "setting sun," which is the attitude of fear, cowardice, and trying to secure ourselves against death.

The setting-sun world of darkness, degrading and belittling human nature, tries to reduce men and women to being only apes or machines. Setting-sun mentality is universal: trying to find a safe home, keeping the hungry neighbor at bay by stimulating an appropriate level of fear, taking refuge in family, hometown, and national pride, and hoping that someone else will do something. It is precisely the world that psychiatrist Walsh describes in which individual defense mechanisms, fear, and narrow beliefs become social norms. This world is devoid of heart. It is without that sympathy, kindness, and warmth toward others which is the binding factor of all human societies.

Human nature is degraded by the refusal to acknowledge the authenticity of men and women who have actually tried to give up the mentality of always seeking personal comfort and security in order to be genuinely who they are. Thus, any sense of the possibility of genuine leadership and genuine wisdom is lost in the urge to keep everything at a level where we can all be comfortable. There is no force of evil in the world which we can blame for such degradation. The setting-sun mentality is one of being lazy and frightened. One is afraid to live fully a life which is transitory, unpredictable, and full of power and natural beauty, a life guided by the understanding symbolized by the Great Eastern Sun.

In the Great Eastern Sun image is the recognition that we do in fact have the power to uplift our lives, to cheer up and celebrate. The energy to uplift our lives and enjoy our world is always available and is the basis of our existence. The Great Eastern Sun is primordial, beyond any particular history or any particular culture. Although it has been pointed out many times, it is beyond any particular teacher or teachings. Wherever humans have been, on any continent, in any epoch, there have always been expressions of great harmony and dignity, joy and celebration, kindness and caring for others. To bring about such a society is the inherent capability of human beings, which is shown again and again no matter what the apparent obstacles. This is why it is called "great."

"East" is the direction of going forward. It is not geographical but it is whichever direction one faces, so long as one goes directly forward to meet the challenge and obstacles in one's life, rather than retreating from them. East is the direction of dawn and awakening. The discovery of goodness dawns in us afresh, and one's own life and the lives of others are warmed and enriched by that sudden glimpse of sacredness which spontaneously arises and is constantly available.

Finally, "Sun" represents brilliance and sharpness. It is an unconditioned view which spans petty mind, cowardice, and the darkness of confusion, and therefore accommodates and illuminates all the smallest details of mind, body, and society. The brilliance of an unconditioned perception shows us what direction to take at each step of our life, gives us discipline, and guides us in what to accept as worthwhile and what to avoid as harmful.

According to the teachings of Shambhala, in every culture, every institution, and every individual, and also in our own minds, we can find both attitudes. The path of warriorship is to begin to discriminate between them and gradually to open again and again into the Great Eastern Sun world through the practice of meditation. To step into the Great Eastern Sun world is not to reject the setting sun, but to include and transform it. It is the view of the ordinary world as sacred, from which nothing is excluded. Everything has its place, even the most seemingly demeaning situation. In the setting-sun world everything appears as meaningless, humorless, and spoiled. Great Eastern Sun vision gently clears up the mess of disharmony as it reawakens the sense of basic unconditioned goodness.

According to Trungpa, the vision of the Great Eastern Sun is not simply an intellectual vision, it brings an entirely new perception of the world: "When you are fully gentle without arrogance you see the brilliance of the universe. You develop a true perception of the universe." He is talking here about our ordinary five sense, as well as the thought process. How we see colors, how we hear sounds, and so on, all of this is transformed by Great Eastern Sun vision so that we begin to experience this world as sacred.

"The idea of sacred world is that, although you see the confusions and problems that fill the world, you also see that phenomenal existence is constantly being influenced by the vision of the Great Eastern Sun. You see that there is a potential for sacredness in every situation. The sacred world begins to show you how you are woven together with the richness and brilliance of the phenomenal world."

Dralas: Wisdom and Power of the Ordinary World

Such sacredness is inherent in the structure of the world. It does not come from outside, nor does it need to be thought up or planned for. In the previous chapters we saw that all experience arises spontaneously, moment by moment, out of an unconditioned ground of goodness. We saw that there is each moment the possibility that we may recognize the vastness of our perceptions. In this way we can begin to connect ourselves to the tremendous resources of power and energy which are in the phenomenal world. We usually exclude ourselves from this connection: we cannot manipulate it nor control it for our personal benefit, and we fear that it will overwhelm us. We have discussed in previous chapters the fundamental fear which usually directs and limits our perception. Through the practice of mindfulness and awareness we begin to recognize and touch this fear, thus perhaps transforming it into fearlessness. Then it is no longer a barrier separating us from the brilliant, vivid, and crystal-clear ordinary world, which is beyond separateness and fragmentation, beyond being for us or against us. When we are able even for a moment to overcome fear in this way, we find that resources of energy and intuitive wisdom are available to us. Trungpa calls these resources "*dralas,*" or "energies beyond aggression."

Dralas are forces or embodiments of power in the phenomenal world which can be directly experienced in the human organism when we are able to tune our perception to them. They are not "supernatural," but are larger patterns of coherent energy and relationships than those which are normally catalysts of human perception. We have discussed such larger patterns of mental process in our study of Bateson's work. As our perception broadens and opens through training, we may begin to connect with the world at a deeper level than mere superficial consciousness. At this deeper level, the dralas can be experienced.

Chögyam Trungpa describes dralas in this way: "Drala could almost be called an entity. It is not quite on the level of a god or gods, but it is an individual strength that does exist. Therefore, we not only speak of drala principle, but we speak of meeting the 'dralas.' The dralas are the elements of reality—water of water, fire of fire, earth of earth—anything that connects you with the elemental quality of reality, anything that reminds you of the depth of perception. There are dralas in the rocks or the trees or the mountains or a snowflake or a clod of dirt. Whatever is there, whatever you come across in your life, those are the dralas of reality. When you make that connection with the elemental quality of the world, you are meeting dralas on the spot; at that point, you are meeting them. That is the basic existence human beings are all capable of. We always have possibilities of discovering magic. Whether it is medieval times or the twentieth century, the possibility of magic is always there.

"A particular example of meeting drala, in my personal experience, is flower arranging. Whatever branches you find, none of them is rejected as ugly. They can always be included. You have to learn to see their place in the situation; that is the key point. So you never reject anything. That is how to make a connection with the dralas of reality.

"Drala energy is like the sun. If you look in the sky, the sun is there. By looking at it, you don't produce a new sun. You may feel that you created or made today's sun by looking at it, but the sun is eternally there. When you discover the sun in the sky, you begin to communicate with it. Your eyes begin to relate with the light of the sun. In the same way, drala principle is always there. Whether you care to communicate with it or not, the magical strength and wisdom of reality are always there . . . . By relaxing the mind, you can reconnect with that primordial, original ground, which is completely pure and simple. Out of that, through the medium of your perceptions, you can discover magic, or drala. You actually can connect your own intrinsic wisdom with a sense of greater wisdom or vision beyond you.

"You might think that something extraordinary will happen to you when you discover magic. Something extra-ordinary does happen. You simply find yourself in the realm of utter reality, complete and thorough reality."

The term "drala" literally means "above" (*la*) "enemy" (*dra*), that is, above aggression. It derives from the Bön tradition, the native religion of Tibet which preceded Buddhism. Bön was a pantheistic tradition that included ritual practices directed toward making a relationship with natural forces, embodied in the dralas. One finds very similar beliefs and practices in almost all societies before they adopted the veneer of scientific belief, thus cutting themselves off from their natural roots. As well as the Shinto religion of Japan, the Native American religions contain this understanding of the interweaving of mind and nature, particularly in their attitude to their environment and their understanding of sacred ground and sacred place. There were also, almost certainly, elements of this in the Greek and Roman traditions. As Trungpa says: "The Greeks and the Romans laid out their cities with some understanding of external drala. You might say that putting a fountain in the center of a square or at a crossroads is a random choice. But when you come upon that fountain, it does not feel random at all. It is in its own place and it seems to enhance the space around it. In modern times, we don't think very highly of the Romans. With all of their debauchery and corrupt rulers, we tend to downplay the wisdom of their culture. Certainly, corruption dispels drala. But there was some power and wisdom in the Roman civilization, which we should not overlook."

This understanding of a deeper wisdom that includes—but goes beyond—rationality continued at least into the early Middle Ages in Europe. Thomas Goldstein elegantly documents the interaction between mystical and scientific insights during this period in *Dawn of Modern Science*: "Sufficient uncertainty surrounds the ultimate philosophical premises to admit of at least a reasonable possibility that those invisible (and often unverifiable) forces may in fact exist and exert their mysterious influences on our everyday environment. Patently modern science has come around to accept some curious phenomena that the more literal-minded nineteenth century scientists would have laughed off as humbug but Medieval mystics might have accommodated without any difficulty at all: the manifestation of extrasensory perception; the varied signals and intrusions of the unconscious of modern psychoanalysis (perhaps even the 'collective unconscious' of Carl Gustav Jung); or, on a somewhat different plane, modern scientists' growing suspicion that our customary premises and modes of thought may break down before certain, as yet unexplored physical phenomena—whether subatomic or supragalactic—conceivably signaling basic deficiencies in our methodological approach. Medieval mystics took it for granted that unknown forces are acting on us from somewhere outside (or indeed inside) ourselves; they would scarcely have been surprised to find our meticulous pragmatic rationalism faltering before a deeper comprehension of the world. Perhaps these and a thousand other symptoms, may suggest that the pragmatic rationalism (or 'rationist positivism') of a more confident phase of our modern age has run its course, that Medieval mysticism may well have contained some grains of a higher wisdom—besides a rich dimension of the experience—that has been essentially closed to us ever since. In science at any rate, the Medieval mind worked in a way that was able to combine the mystic with the pragmatic approach, untested magic beliefs with straight empirical observation."[7]

Let us look a little further into the drala principle, or "kami," as it is known in the Japanese Shinto tradition. Shinto, the way of the kami, is a way of human life, "an amalgam of attitudes, ideas, and ways of doing things that through two millenia and more have become an integral part of the way of the Japanese people," according to Sokyo Ono, a Japanese authority on Shinto. He describes the term "kami" as an honorific term for noble, sacred spirits which implies a sense of respect, love and awe. He says, "All beings have such spirits so in a sense all beings can be called kami or regarded as potential kami."

Ono describes kami very clearly. I will quote his description at length: "Among the objects or phenomena designated as kami are the qualities of growth, fertility and production; natural phenomena such as wind and thunder; natural objects such as the sun, mountains, rivers, trees and rocks; some animals; and ancestral spirits . . . . Also regarded as kami are the guardian spirits of the land, occupations, and skills; the spirits of national heroes, men of outstanding deeds or virtues, and those who have contributed to civilisation, culture and human welfare; those who have died for the state or the community and the pitiable dead. . .

"It is true that in many instances there are kami which apparently cannot be distinguished from the deities and spirits of animism or animatism, but in modern Shinto all kami are conceived in a refined sense to be spirits with nobility and authority. The kami concept today includes the idea of justice, order and divine favour (blessing), and implies the basic principle that the kami function harmoniously in cooperation with one another and rejoice in the evidence of harmony and cooperation in this world.

"In Shinto there is no absolute deity that is the creator and ruler of all. The creative function of the world is realized through the harmonious cooperation of the kami in the performance of their respective missions.

"There are many points about the kami concept that cannot be fully understood, and there is disagreement even among modern scholars on this subject. The Japanese people themselves do not have a clear idea regarding the kami. They are aware of the kami intuitively at the depths of their consciousness and communicate with the kami directly without having formed the kami idea conceptually or theologically"[8]

Our urgent need to rediscover a "participating consciousness"— an awareness of our connection with and our participation in a living world, an "enchanted world," is the theme of Morris Berman's *The Reenchantment of the World:* "The view of nature which predominated in the West down to the eve of the Scientific Revolution was that of an enchanted world. Rocks, trees, rivers, and clouds were all seen as wondrous, alive, and human beings felt at home in this environment. The cosmos, in short, was a place of *belonging.* A member of this cosmos was not an alienated observer of it but a direct participant in its drama. His personal destiny was bound up with its destiny, and this relationship gave meaning to his life. This type of consciousness— . . . 'participating consciousness'—involves merger, or identification, with one's surroundings, and bespeaks a psychic wholeness that has long since passed from the scene. Alchemy, as it turns out, was the last great coherent expression of participating consciousness in the West."[9] This "participating consciousness" is, precisely, connecting with drala, or kami, energy. To reconnect with drala energy is to participate with the power of nature rather than to become alienated from it and have to "control" it.

Of course, as Berman points out, this reenchantment cannot take place by our ignoring the insight of science altogether and trying to return to a primitive belief system. It has to be based on a larger awareness that is able to embrace both the rational pragmatism of science and the intuitive wisdom of natural magic and direct perception. These two ways of understanding appear to be polar contrasts and to many have long seemed altogether incompatible. Yet, the ability to accommodate such apparent incompatibles is the stepping stone to greater wisdom. As that brilliant physicist Wolfgang Pauli says, "I can consider the ambition of overcoming opposites, including also a synthesis embracing both rational understanding and the mystical." The apparent conflict between science and intuitive wisdom is perhaps the most profound challenge we have ever had to face. It is the challenge of how to join the brilliance and sharpness of intellect with the sympathetic kindness and direct insight of the heart. To go beyond it would indeed be a genuine step forward for all of humanity.

Although we cannot go very far into it here, there is, of course much to be understood if we are to incorporate understanding of "drala" into our modern technological society. Certainly, as I have argued throughout this book, the two are not incompatible. Indeed, the blending of drala principle with modern outlook may well be the origin of modern Japan's strength. It is clear, however, that the Shambhala Teaching is here reintroducing us to a very powerful aspect of our world which we have long forgotten. It is also important to always bear in mind the nondualistic, nontheistic foundation, the unconditioned nature from within which all of this arises. Gentleness and nonaggression, which come from recognizing this ground, is, according to Trungpa, the only way to connect with drala energy: "But there is still a question as to what it is that allows you to make that connection. The drala principle was likened to the sun. Although the sun is always in the sky, what is it that causes you to look up and see that it is there? Although magic is always available, what allows you to discover it? The basic definition of drala is 'energy beyond aggression.' The only way to contact that energy is to experience a gentle state of being in yourself. So the discovery of drala is not coincidental. To connect with the fundamental magic of reality, there has to be gentleness and openness in you already. Otherwise, there is no way to recognize the energy of nonaggression, the energy of drala, in the world. So the individual training and discipline of the Shambhala warrior are the necessary foundation for experiencing drala."[10]

In this book I have tried to suggest that such a transformation of human existence, the possibility of living together in a society bound by kindness and wakefulness, is not a mere utopian fantasy, philosophical speculation, or religious revelation. The view of the world and how we live in it is profoundly different from the "common sense" view of the twentieth century. However, it is thoroughly grounded in an accurate understanding of the nature of the human body and mind, and of the world in which humanity has its place. There is no mystery to the transformation of perception that I have described, nor to the spiritual path of training that can accomplish it. It is quite ordinary, although it does take energy and warmth, dedication and humor.

Chapter Notes

Chapter One

    1. Mortimer J. Adler, *The Paedeia Proposal* (New York: Macmillan, 1982).
    2. Alvin Toffler, *The Third Wave* (New York: Bantam, 1981); *Future Shock* (New York: Bantam, 1971).
    3. John Naisbitt, *Megatrends* (New York: Warner, 1982).
    4. Marilyn Ferguson, *The Aquarian Conspiracy* (Los Angeles: J.P. Tarcher, 1980).
    5. Ramana Maharshi, *The Collected Work of Ramana Maharshi*. Edited by Arthur Osborne. (London: Rider, 1968); G. I. Gurdjieff, *In Search of the Miraculous* (New York: Harcourt Brace Jovanovich, 1965); Krishnamurti, *Talks and Dialogues* (New York: Avon, 1970); Chögyam Trungpa, *Shambhala: The Sacred Path of the Warrior* (Boulder, Co.: Shambhala Publications, 1984).
    6. Trungpa, *Shambhala: The Sacred Path of the Warrior.*
    7. Guy R. Welbon, *Buddhist Nirvana and its Western Interpreters* (Chicago: University of Chicago Press, 1968). Rick Fields, *How the Swans Came to the Lake* (Boulder, Co.: Shambhala Publications, 1982).
    8. Calvin Tomkins, *The Bride and the Bachelors* (New York: Penguin, 1976).
    9. David Yankelovitch, *New Rules* (New York: Bantam, 1982).
    10. William James, *Psychology: A Brief Course* (New York: Dover, 1961).
    11. Daniel Goleman, *The Varieties of Meditative Experience* (New York: E. P. Dutton, 1977).
    12. Kenneth Pelletier, *Towards a Science of Consciousness* (New York: Delacorte, 1978).

Chapter Two

    1. Alfred North Whitehead, *Science and the Modern World* (New York: Free Press, 1967).

2. The Sanskrit terms for the three natures (*trisvabhava*) are (a) *parinispanna,* (b) *paratantra,* and (c) *parikalpita.* Helpful discussions of these are found in Herbert Guenther, *Buddhist Philosophy in Theory and Practice* (New York: Penguin, 1972): Janice Dean Willis, *On Knowing Reality* (New York: Columbia University Press, 1979); and Thomas Kochumulton, *A Buddhist Doctrine of Experience* (Delhi: Motilal Banarsidas, 1982).

The interpretation of the three natures we have presented here is the original view of the Indian founders of the *yogacara* tradition, Asanga and Vasubandhu, a view that was adopted by the Tibetan Vajrayana schools; it is used in Guenther and Willis. The Chinese interpretation, given by Kochumulton, differs in that a more definite separation is made between *parinispanna* and *paratantra. Paratantra* is aligned more with *parikalpita.* In this interpretation, which was peculiarly Chinese but which passed on to Japan, there is a possibility of beginning to regard *parinispanna,* the absolute truth, as altogether transcendent or "otherworldly."

3. On eternalism and nihilism in Buddhist philosophy see, for example, Ramanan, *Nagarjuna's Philosophy* (Bombay: Bhavatya Vidya Prakasha, 1971).

4. Willis, *On Knowing Reality.*

5. Alfred Korzybski, selections from *Science and Sanity* (International non-Aristotelian Library, 1972).

6. George Spencer-Brown, *Laws of Form* (New York: Bantam, 1973).

7. David Bohm, *Wholeness and the Implicate Order* (London: Routledge and Kegan Paul, 1983).

8. Guenther, *Buddhist Philosophy in Theory and Practice.*

9. Herbert Guenther, *Tibetan Buddhism in Western Perspective* (Berkeley: Dharma Publishing, 1977).

10. Trungpa, *Shambhala: The Sacred Path of the Warrior.*

Chapter Three

1. Quoted in Earl Miner, *An Introduction to Japanese Court Poetry* (Palo Alto, Ca.: Stanford University Press, 1968).

2. For an introduction to linguistic philosophy, see Richard Rorty, ed., *The Linguistic Turn* (Chicago: University of Chicago Press, 1967).

3. Quoted in William Barrett, *The Illusion of Technique* (New York: Anchor Press/Doubleday, 1978).

4. Harold Brown, *Perception, Theory and Commitment* (Chicago: University of Chicago Press, 1977).

5. Karl Popper, *Objective Knowledge* (Oxford: Oxford University Press, 1977).

6. John Lyons, *Semantics I* (Cambridge: Cambridge University Press, 1977).

7. George Lakoff and Marle Johnson, *Metaphors We Live By* (Chicago: University of Chicago Press, 1980).

8. F. Palmer, *Semantics*, second edition (Cambridge: Cambridge University Press, 1981).

9. Alton Becker, *Text-Building, Epistemology and Aesthetics in Javanese Shadow Theater* (unpublished manuscript, 1977).

10. For good basic introductions to the various schools of linguistics, see John Lyons, *Language and Linguistics* (Cambridge: Cambridge University Press, 1981); G. Sampson, *Schools of Linguistics* (Palo Alto: Stanford University Press, 1980); and Palmer, *Semantics*. For a more advanced discussion of semantics, see Lyons, *Semantics I*. Much of the following discussion is based on material in these references.

11. Quoted in Sampson, *Schools of Linguistics*.

12. Benjamin Whorf, *Language, Thought and Reality* (Cambridge, Ma.: MIT Press, 1956).

13. For a discussion of Chomsky's work which is not too technical, see his *Language and Responsibility* (New York: Pantheon, 1979).

14. For a discussion of Popper's three worlds, see especially his *Objective Knowledge: The Open Universe* (London: Rowen and Littlefield, 1982); and, with Eccles, *The Self and Its Brain* (New York: Springer International, 1981).

15. Popper, *Objective Knowledge*.

16. Lakoff and Johnson, *Metaphors We Live By*.

17. Miner, *An Introduction to Japanese Court Poetry*.

18. Martin Heidegger, *What Is Called Thinking* (New York: Harper and Row, 1968).

Chapter Four

1. Thomas Goldstein, *Dawn of Modern Science* (Boston: Houghton Mifflin, 1980).

2. F. Heer, *The Medieval World* (New York: New American Library/Mentor, 1964).

3. Kenneth Clark, *Civilization* (New York: Harper and Row, 1970).

4. Herbert Butterfield, *The Origins of Modern Science* (New York: Free Press, 1965).

5. For good introductions to the origins of modern science, see Goldstein, *Dawn of Modern Science;* Whitehead, *Science and the Modern World;* Butterfield, *The Origins of Modern Science;* R. G. Collingwood, *The Idea of Nature* (Oxford: Oxford University Press, 1960); and Charles Singer, *A Short History of Scientific Ideas* (Oxford: Oxford University Press, 1959).

6. Frances Yates, *Giordano Bruno and the Hermetic Tradition* (Chicago: University of Chicago Press, 1964).

7. Morris Berman, *The Reenchantment of the World* (Ithaca, N.Y.: Cornell University Press, 1981).

8. Heinz Pagels, *The Cosmic Code* (New York: Bantam, 1983).

Chapter Five

1. Paul Davies, *Other Worlds* (New York: Simon and Schuster, 1980).

2. Heer, *The Medieval World.*

3. *Ibid.*

4. T. J. Kaptchuk, *The Web That Has No Weaver* (New York: Congdon and Weed, 1983).

5. R. Katz, *Boiling Energy: Community Healing Among the Kalahari Kung* (Cambridge, Ma.: Harvard University Press, 1982).

6. M. McCloskey, "Intuitive Physics," *Scientific American* (April, 1983).

7. Fritjof Capra, *The Turning Point* (New York: Simon and Schuster, 1982).

8. Karl Popper, *Logic of Scientific Discovery* (New York: Harper and Row, 1959); *Conjectures and Refutations* (London: Routledge and Kegan Paul, 1963); and *Objective Knowledge.*

9. Karl Popper, *Quantum Mechanics and the Schism in Physics* (London: Hutchinson, 1982).

10. Popper, *Objective Knowledge.*

11. For simple, concise descriptions of this decisive step, see Brown, *Perception, Theory and Commitment;* A. F. Chalmers, *What is This Thing Called Science* (Queensland: University of Queensland Press, 1982). For a more in-depth and technical presentation, see F. Suppe, *The Structure of Scientific Theories* (Champaign, Ill.: University of Illinois Press, 1974). Some of the key works contributing to this

revolution are Thomas Kuhn, *The Structure of Scientific Revolutions* (Chicago: University of Chicago Press, 1962); Stephen Toulmin, *Foresight and Understanding* (London: Hutchinson, 1963); Norwood Hanson, *Patterns of Discovery* (Cambridge: Cambridge University Press, 1958); and Paul Feyerabend, *Against Method* (London: Verso, 1978).

12. Hans Eysenck and Carl Sargent, *Explaining the Unexplained* (London: Weidenfeld and Nicolson, 1982).

13. J. B. Priestley, *Man and Time* (London: Aldus Books, 1964).

14. Alex Comfort, *Reality and Empathy* (Albany, N.Y.: State University of New York, 1984).

15. Michael Shallis, *On Time* (New York: Schocken, 1983).

16. Popper, *Objective Knowledge*. For arguments rejecting precognition as valid observational data, see, for example, John Taylor, *Science and the Supernatural* (New York: E. P. Dutton, 1980); and George Abell and Barry Singer, *Science and the Paranormal* (New York: Scribners, 1981).

Chapter Six

1. Charles Darwin, *The Origin of Species* (New American Library/Mentor, 1958).

2. Butterfield, *The Origin of Modern Science*.

3. Darwin, *The Origin of Species*.

4. Herbert Spencer, *Social Statics* (London?: Chapman, 1851).

5. Thomas Huxley, "The Struggle for Existence in Human Society," in *The Nineteenth Century* (February, 1888).

6. Quoted in Jeremy Rifkin, *Algeny* (New York: Viking, 1983).

7. Richard Dawkins, *The Selfish Gene* (Oxford: Oxford University Press, 1976).

8. Pierre Grassé, *Evolution of Living Organisms* (New York: Academic Press, 1977).

9. Popper, *Objective Knowledge*.

10. Leonard Matthews, introduction to *The Origin of Species* (London: Dent and Son, 1971).

11. Darwin, *The Origin of Species*.

12. Herbert Nilsson, *Synthetische Arbildung* (Stockholm: Lund University, 1954).

13. Francis Hitching, *The Neck of the Giraffe* (New York: New American Library/Mentor, 1982).

14. Melvin Konner, *The Tangled Wing* (New York: Holt, Rinehart and Winston, 1982).

15. Robert Axelrod and Wilber Hamilton in John Maynard Smith, ed., *Evolution Now* (San Francisco: Freeman, 1983).

16. George Ledyard Stebbins, *Darwin to DNA, Molecules to Men* (San Francisco: Freeman, 1982).

Chapter Seven

1. Petr Kropotkin, *Mutual Aid* (London: Porter Sargent, 1954).

2. John A. Thompson and Patrick G. Geddes, *Life: Outlines of General Biology*, quoted in Rifkin, *Algeny*.

3. Edward O. Wilson, *Sociobiology* (Cambridge, Ma.: Harvard University Press, 1975).

4. Dawkins, *The Selfish Gene*.

5. John Tyler Bonner, *The Evolution of Culture in Animals* (Princeton, N.J.: Princeton University Press, 1980).

6. Edward O. Wilson and Charles J. Lumsden, *Promethean Fire* (Cambridge, Ma.: Harvard University Press, 1983).

7. Edward O. Wilson, *On Human Nature* (Cambridge, Ma.: Harvard University Press, 1978); Georg Breuer, *Sociobiology and the Human Dimension* (Cambridge: Cambridge University Press, 1982); Mary Midgley, *Beast and Man* (Ithaca: Cornell University Press, 1982); and Gunther S. Stent, ed., *Morality as a Biological Phenomenon* (Berkeley: University of California Press, 1980).

8. Konner, *The Tangled Wing*.

9. H. B. Barlow, "Nature's Joke: A Conjecture on the Biological Role of Consciousness," and N. K. Humphrey, "Nature's Psychologists," in Josephson and Ramachandran, eds., *Consciousness and the Physical World* (Elmsford, N.Y.: Pergamon Press, 1980).

10. Wilson, *On Human Nature*.

Chapter Eight

1. Gordon Rattray Taylor, *The Great Evolution Mystery* (New York: Harper and Row, 1983). For a review of the new-Darwinist response to some of these criticisms, see John Maynard Smith, ed., *Evolution Now* (San Francisco: Freeman, 1982).

2. Taylor, *ibid.*

3. *Ibid.*

4. Stebbins, *Darwin to DNA, Molecules to Men*.

5. Ilya Prigogine, *From Being to Becoming: Time and Complexity in the Physical Sciences* (San Francisco: Freeman, 1980); *Order Out of Chaos: Man's New Dialogue With Nature* (Boulder: New Science Library, 1984).

6. Erich Jantsch, *The Self-Organizing Universe* (Elmsford, N.Y.: Pergamon, 1980).

7. Gerald Feinberg and Robert Shapiro, *Life Beyond Earth* (New York: William Morrow, 1980).

8. James E. Lovelock, *Gaia: A New Look at Life on Earth* (Oxford: Oxford University Press, 1979).

9. Humberto Maturana and Francisco Varela, *Autopoesis and Cognition* (Dordrecht, The Netherlands: Reidel, 1979).

10. *Ibid.*

11. Francisco Varela, *Principles of Biological Autonomy* (Amsterdam: North Holland, 1979).

12. Gregory Bateson, *Mind and Nature, A Necessary Unity* (New York: Bantam, 1980).

Chapter Nine

1. There are many good introductory books on relativity. Simple and easy reading are Herman Bondi, *Relativity and Common Sense* (New York: Dover, 1980); Nigel Calder, *Einstein's Universe* (New York: Penguin, 1980); and Albert Einstein and Leopold Infeld, *The Evolution of Physics* (New York: Simon and Schuster, 193?).

2. This particular version of the well-known Twin Paradox is adapted from Shallis, *On Time*.

3. Quoted in Gerald Holton, *Thematic Origins of Scientific Thought* (Cambridge, Ma.: Harvard University Press, 1973).

4. Prigogine, *Order Out of Chaos*.

5. Shallis, *On Time*.

6. *Ibid.*

7. Becker, *Text-building, Epistemology and Aesthetics in Javanese Shadow Theatre*.

8. Quoted in Carl G. Jung, *Synchronicity* (Princeton, N.J.: Princeton University Press, 1973).

9. Helmut Wilhelm, *Heaven, Earth and Man in the Book of Changes* (Seattle: University of Washington Press, 1979).

10. Richard Wilhelm, *Lectures on the I Ching: Constancy and Change* (Princeton, N.J.: Princeton University Press, 1979).

11. Davies, *Other Worlds*.

12. Patrick Heelan, *Space Perception and the Philosophy of Science* (Berkeley: University of California Press, 1983).

Chapter Ten

1. There are several excellent nontechnical books on elementary particles and quantum mechanics, the topics of this and the next chapters. The best are Pagels, *The Cosmic Code*; Davies, *Other Worlds*; and John C. Polkinghorne, *The Particle Play* (San Francisco: Freeman, 1981).

2. Bohm, *Wholeness and the Implicate Order*.

3. This very imaginative and clear version of the famous "double-slit experiment" is due to Richard Feynman and is described in Pagels's *The Cosmic Code*.

4. Bernard d'Espagnat, "Quantum Theory and Reality," in *Scientific American* (November, 1979).

5. Described in João Andrade e Silve and Georges Lochak, *Quanta* (London: World University Library, 1969).

6. Quoted in Pagels, *The Cosmic Code*.

Chapter Eleven

1. The best reviews of the various interpretations of quantum mechanics are in Heinze Pagels, *The Quantum Code* (New York: Simon and Schuster, 1982); and Davies, *Other Worlds*. An excellent but more technical review is Bernard d'Espagnat, *The Conceptual Foundations of Quantum Mechanics* (London: Benjamin, 1971). For some of the early key papers on the topic, see Stephen Toulmin, ed., *Physical Reality* (New York: Harper and Row, 1970).

2. Niels Bohr, *Atomic Physics and Human Knowledge* (Science Editions, 1958); *Essays 1958-1962* (Interscience, 1963); and Ruth Moore, *Niels Bohr* (New York: Knopf, 1966).

3. Max Jammer, *Conceptual Development of Quantum Mechanics* (New York: McGraw Hill, 1966).

4. William James, *The Principles of Psychology* (New York: Dover, 1950), quoted in Holton, *Thematic Origins of Scientific Thought*.

5. Quoted in Holton, *Thematic Origins of Scientific Thought*.

6. Niels Bohr, "Quantum Physics and Philosophy," in *Essays 1958-1962*.

7. Quoted in Holton, *Thematic Origins of Scientific Thought*.

8. Werner Heisenberg, *Physics and Philosophy* (New York: Harper and Row, 1958).

9. Popper, *The Open Universe;* and *Quantum Theory and the Schism in Physics.*

10. Heisenberg, *Physics and Philosophy.*

11. Popper and Eccles, *The Self and Its Brain.*

12. John Eccles, quoted in Paul Davies, *God and the New Physics* (New York: Simon and Schuster, 1983).

13. Eugene Wigner, *Symmetries and Reflections* (Cambridge, Ma.: MIT Press, 1970).

14. Evans Harris Walker, in A. Puhanak, ed., *The Iceland Papers* (Essentia Association, 1979).

15. Bryce deWitt and Neil Graham, *The Many Worlds Interpretation of Quantum Mechanics* (Princeton, N.J.: Princeton University Press, 1973).

16. Davies, *Other Worlds; The Accidental Universe* (Cambridge: Cambridge University Press, 1982).

17. Bohm, *Wholeness and the Implicate Order.*

18. Ken Wilber, ed., *The Holographic Paradigm and Other Paradoxes* (Boulder, Co.: Shambhala Publications, 1982).

19. Comfort, *Reality and Empathy.*

Chapter Twelve

1. Popper and Eccles, *The Self and Its Brain.*

2. Paul Maclean, "The Paranoid Streak in Man," in *Beyond Reductionism* (London: Hutchinson, 1969).

3. Charles Hampden-Turner, *Maps of the Mind* (London: Macmillan, 1981). This book provides an excellent summary of Maclean's ideas, on which the following description is based.

4. Quoted in Eccles and Popper, *The Self and Its Brain.*

5. Vernon Mountcastle, "The View from Within," *Johns Hopkins Medical Journal,* #136.

6. Richard L. Gregory, *The Intelligent Eye* (New York: McGraw Hill, 1970); *Mind in Science* (Cambridge: Cambridge University Press, 1981); and in Jonathan Miller, *States of Mind* (New York: Pantheon, 1983).

7. James J. Gibson in R. Shaw and J. Bransford, eds., *Perceiving, Acting, Knowing* (New York: Wiley, 1977).

8. John Bowlby, *Attachment and Loss, Vol. 1* (London: Hogarth Press, 1969).

9. Donald O. Hebb, "On the Nature of Fear," *Psychological Review* 53, 1946.

10. Konner, *The Tangled Wing*.

11. See in Eccles and Popper, *The Self and Its Brain*.

12. Wilder Penfield, *The Mystery of the Mind* (Princeton, N.J.: Princeton University Press, 1975).

13. L. Weiskrantz *et al.*, "Blindsight," *The Lancet* (April 1974).

Chapter Thirteen

1. Francisco Varela, "Living Ways of Sense-Making," paper read at the International Symposium on Order and Disorder, Stanford University, 1981.

2. Bateson, *Mind and Nature, a Necessary Unity*.

3. Gregory Bateson, *Steps to an Ecology of Mind* (New York: Ballantine, 1975).

4. *Ibid*.

5. *Ibid*.

6. Edwin Land, "Our 'Polar Partnership' with the World Around Us," in *Harvard Magazine*, 1978.

7. Davies, *God and the New Physics*.

8. Victor Frankl, quoted in Charles Hampden-Turner, *Maps of the Mind*.

Chapter Fourteen

1. For a simple, good introduction to process philosophy, see Ivor Leclerc, *Whitehead's Metaphysics* (London: Allen and Unwin, 1958); Victor Lowe, *Understanding Whitehead* (Baltimore: Johns Hopkins University Press, 1966). The primary exposition of process philosophy is Alfred North Whitehead, *Process and Reality* (New York: Free Press, 1969). A helpful introduction and key to this rather difficult book is Donald W. Sherburne, *A Key to Whitehead's Process and Reality* (Bloomington, Ind.: Indiana University Press, 1966). Other works of Whitehead from his later period, dealing with aspects of process philosophy, are *Science and the Modern World; Adventures of Ideas* (New York: Free Press, 1967); *Symbolism* (New York: Putnam, 1959); and *Modes of Thought* (New York: Free Press, 1968).

2. For Whitehead's discussion of religion, see *Religion in the Making* (London: Macmillan, 1930). For a discussion of process philosophy and Buddhism, see *Philosophy East and West*, Volume XXV, no. 4 (October 1975). This volume is entirely devoted to a conference

on Mahayana Buddhism and Whitehead, held at the University of Hawaii, November, 1974.

3. There is very little biographical material. A sense of his style is conveyed delightfully in Lucien Price, *Dialogues of Alfred North Whitehead* (Boston: Little Brown, 1954). A short autobiographical sketch is in Alfred North Whitehead, *Essays in Science and Philosophy* (Philosophical Library, 1947). This volume also contains essays on education, a topic to which Whitehead returned again and again throughout his life. Other writing on education is in Alfred North Whitehead, *The Aims of Education* (Tonbridge, U. K.: Ernest Benn, 1966).

4. Whitehead, *Process and Reality*, p. 9.

5. Whitehead, *Adventure of Ideas*, p. 133.

6. Leclerc, *Whitehead's Metaphysics*.

7. Whitehead, *Modes of Thought*, p. 111.

8. *Ibid.*, p. 112.

9. *Ibid.*, p. 116.

10. Leclerc, *Whitehead's Metaphysics*.

11. Whitehead, *Process and Reality*, p. 204.

12. *Ibid.*, p. 207.

13. Sherburne, *A Key to Whitehead's Process and Reality*.

14. Whitehead, *Process and Reality*, p. 108.

15. *Ibid.*, p. 109.

16. *Ibid.*

17. Robert Palter, *Whitehead's Philosophy of Science* (Chicago: University of Chicago Press, 1960).

18. Rupert Sheldrake, *A New Science of Life* (Los Angeles: J. P. Tarcher, 1981).

19. Whitehead, *Process and Reality*, p. 407.

20. Whitehead, *Adventure of Ideas*, p. 285.

Chapter Fifteen

1. Much of the material in this chapter, as well as in Chapter 16, comes from the oral teachings of Chögyam Trungpa. Each year, Trungpa meets with approximately three hundred of his students in a three-month retreat for the practice and study of the Buddhist Mahayana and Vajrayana teachings—the Vajradhatu Seminary. During these retreats, for the past ten years, Trungpa has given series of lectures based on traditional texts, particularly those of the Rimé tradition of Tibetan Buddhism gathered by the nineteenth-century teacher, Jamgön Kongtrül the Great. These lectures have been

transcribed and published in limited editions, available to authorized students. Portions of the transcripts are now being prepared for general publication. Other sources in which some of the material of this chapter may be found are Trungpa, *Glimpses of Abhidharma* (Boulder, Co.: Shambhala Publications, 1977); *Cutting Through Spiritual Materialism* (Boulder, Co.: Shambhala Publications, 1973); *Shambhala: The Sacred Path of the Warrior;* and Late Rinbochay and E. Napper, *Mind in Tibetan Buddhism* (Valois, N.Y.: Gabriel, 1980); Herbert Guenther, *Kindly Bent to Ease Us* (Berkeley: Dharma Publishing, 1975 and 1976); Geshe Rabten, *The Mind and Its Functions,* Gelong Tubkay, tr. (Switzerland: Tharpa Choeling, 1978).

    2. Edward Conze, *Buddhist Thought in India* (Ann Arbor: University of Michigan Press, 1970).

    3. Trungpa, *Shambhala: The Sacred Path of the Warrior.*

    4. Trungpa, *Vajradhatu Seminary Transcripts, 1979* (Boulder, Co.: Vajradhatu Publications, 1980).

    5. Trungpa, *Cutting Through Spiritual Materialism;* Conze, *Buddhist Thought in India.*

    6. Guenther, *Buddhist Philosophy in Theory and Practice.* We are particularly following the development as it is described by the practicing lineages of the Kagyü and Nyingma schools.

    7. Yensho Kanakura, "Hindu and Buddhist Thought in India," *Hokke Journal* (Tokyo: Yokohama, 1980).

    8. D. T. Suzuki, *The Lankavatara Sutra* (Boulder, Co.: Shambhala Publications, 1982).

    9. Ramana, *Nagarjuna's Philosophy.* For the Madhyamika criticism of the earlier stages, see Geshe Kelsang Gyatso, *Meaningful to Behold* (Chapter 9) (Ulverston, U.K.: Wisdom Publications, 1980).

    10. Guenther, *Tibetan Buddhism in Western Perspective.*

    11. Trungpa, *Cutting Through Spiritual Materialism.*

    12. Trungpa, *Shambhala: The Sacred Path of the Warrior.*

    13. *Ibid.*

Chapter Sixteen

    1. Quoted in William H. Shannon, *Thomas Merton's Dark Path* (New York: Penguin, 1982).

    2. Bernadette Roberts, *The Experience of No-Self* (Boulder, Co.: Shambhala Publications, 1984).

    3. Duane Elgin, *Voluntary Simplicity* (New York: Morrow, 1981).

4. Chögyam Trungpa, *Garuda IV: The Foundations of Mindfulness* (Boulder, Co.: Shambhala Publications, 1975).

5. *Ibid.*, and Nyanaponika Thera, *The Heart of Buddhist Meditation* (New York: Samuel Weiser, 1979).

6. Herbert Guenther and Leslie Kawamura, *Mind in Buddhist Psychology* (Berkeley: Dharma Publishing, 1975).

7. *Ibid.*

8. Gustie L. Herrigel, *Zen in the Art of Flower Arranging* (London: Routledge & Kegan Paul, 1958).

9. Chögyam Trungpa, *Kalapa Ikebana Newsletter*, Winter 1984.

10. Trungpa, *Shambhala: The Sacred Path of the Warrior*.

Chapter Seventeen

1. Roger Walsh, *Staying Alive: The Psychology of Human Survival* (Boulder, Co.: Shambhala Publications, 1984).

2. Three good introductory books on the Buddhist tradition are Ösel Tendzin, *Buddha in the Palm of Your Hand* (Boulder, Co.: Shambhala Publications, 1982); Walpola Rahula, *What the Buddha Taught* (London: Gordon Frazer, 1978); and Richard H. Robinson, *The Buddhist Religion* (Dickenson, 1970). The first of these is written from the point of view of the practitioner, the second is a simple scholarly statement of the basic doctrines, and the third is historical.

3. A very simple but extraordinarily insightful introduction to Confucian principles is Herbert Finagarette, *Confucianism: The Secular as Sacred* (New York: Harper and Row, 1972). An excellent introduction to the role of Confucianism in Chinese society, government, and arts is Charles O. Hucker, *China's Imperial Past* (Palo Alto, Ca.: Stanford University Press, 1975). Another simple introduction to Confucius is Raymond Dawson, *Confucius* (New York: Hill and Wang, 1981).

4. Mencius, quoted in *A Source Book in Chinese Philosophy*, Wing-Tsit Chang, ed. (Princeton, N.J.: Princeton University Press, 1963).

5. Burton Watson, tr., *Hsun Tzu: Basic Writings* (New York: University Press, 1963).

6. Quoted in Joseph Neeham and Colin Ronan, *The Shorter Science and Civilisation in China*, Vol. I (Cambridge: Cambridge University Press, 1978).

7. Wh. Theodore Bary, *The Unfolding of Neo-Confucianism* (New York: Columbia University Press, 1970).

8. Hucker, *China's Imperial Past.*

9. Wilhelm, *Heaven, Earth and Man in the Book of Changes;* Chan, *A Source Book in Chinese Philosophy;* and Trungpa, *Shambhala: The Sacred Path of the Warrior.*

10. Hucker, *China's Imperial Past.*

Chapter Eighteen

1. Ken Wilber, *Eye to Eye* (New York: Doubleday/Anchor, 1983).

2. Alexandra David-Neel, *Initiations and Initiates in Tibet* (Boulder, Co.: Shambhala Publications, 1970).

3. Gurdjieff's own writings are *All and Everything* (New York: E. P. Dutton, 1964); *Meetings with Remarkable Men* (New York: E. P. Dutton, 1967); *Life is Only Real Then, When "I Am"* (New York: E. P. Dutton, 1981).

4. Descriptions of The Fourth Way are to be found in several books by P. D. Ouspensky, particularly, *The Psychology of Man's Possible Evolution* (New York: Vintage, 1974); *The Fourth Way* (New York: Random House, 1971); and *The Search for the Miraculous* (New York: Harcourt Brace Jovanovich, 1965). For commentaries on the teachings of Gurdjieff and Ouspensky as they are practiced in everyday life, see Maurice Nicoll, *Psychological Commentaries on the Teaching of Gurdjieff and Ouspensky*, Volumes I-V (Boulder, Co,: Shambhala Publications, 1984).

5. Chögyam Trungpa, *Born in Tibet* (Boulder, Co.: Shambhala Publications, 1977). Books by Trungpa on Buddhism in the English language are *Meditation in Action* (Boulder, Co.: Shambhala Publications, 1969); *Cutting Through Spiritual Materialism; The Myth of Freedom* (Boulder, Co.: Shambhala Publications, 1976); *Glimpses of Abhidharma;* and *Journey Without Goal* (Boulder, Co.: Shambhala Publications, 1981).

6. The only published work on the teachings of Shambhala to date is Trungpa, *Shambhala: The Sacred Path of the Warrior.*

7. Goldstein, *Dawn of Modern Science.*

8. Sokyo Ono, *Shinto: The Kami Way* (Rutland, Vt.: Charles Tuttle, 1962).

9. Berman, *The Reenchantment of the World.*

10. Trungpa, *Shambhala: The Sacred Path of the Warrior.*

Index

Also in New Science Library

Awakening the Heart: East/West Approaches to Psychotherapy and the Healing Relationship, by John Welwood.

Beyond Illness: Discovering the Experience of Health, by Larry Dossey, M.D.

The Holographic Paradigm and Other Paradoxes, edited by Ken Wilber.

Jungian Analysis, edited by Murray Stein. Introduction by June Singer.

No Boundary: Eastern and Western Approaches to Personal Growth, by Ken Wilber.

Order Out of Chaos: Man's New Dialogue with Nature, by Ilya Prigogine and Isabelle Stengers. Foreword by Alvin Toffler.

Quantum Questions: Mystical Writings of the World's Great Physicists, edited by Ken Wilber.

A Sociable God: Toward a New Understanding of Religion, by Ken Wilber.

Space, Time and Medicine, by Larry Dossey, M.D.

The Sphinx and the Rainbow: Brain, Mind and Future Vision, by David Loye.

Staying Alive: The Psychology of Human Survival, by Roger Walsh, M.D.

The Tao of Physics: An Exploration of the Parallels between Modern Physics and Eastern Mysticism, second edition, revised and updated, by Fritjof Capra.

Up from Eden: A Transpersonal View of Human Evolution, by Ken Wilber.